DATE DUE		
MAY 11	APR 20 '93	
DEC 19 1993	MAY 29 '93	
FEB 13 1984	NOV 29 1993	
APR 27	MAY 21 1994	
OCT 01 1984	OCT 5 1994	
NOV 21 1984	NOV 18 1994	
APR 0 1985	MAY 3	
JUL 8 1986	JAN 2 1996	
JUL 22 1986	FEB 22	
	MAY 19 1997	
MAY 13 1987		
APR 02 1988	JUN 4 1998	
JAN 18 1988		
FEB 22 '89	JAN 18 2000	
DEC 20 '90		
MAR 27 '91		
MAY 7 '91		
MAR 31 '93		

DEMCO 38-297

HAZARDOUS MATERIALS

SCHIELER · PAUZÉ

 VAN NOSTRAND REINHOLD COMPANY
NEW YORK   CINCINNATI   ATLANTA   DALLAS   SAN FRANCISCO
LONDON   TORONTO   MELBOURNE

Van Nostrand Reinhold Company Regional Offices:
New York   Cincinnati   Atlanta   Dallas   San Francisco

Van Nostrand Reinhold Company International Offices:
London   Toronto   Melbourne

Copyright © 1976 by Litton Educational Publishing, Inc.

Library of Congress Catalog Card Number: 76-14448
ISBN 0-442-27394-0

Manufactured in the United States of America

Published by Van Nostrand Reinhold Company
450 West 33rd Street, New York, N.Y. 10001

Published simultaneously in Canada by Van Nostrand Reinhold Ltd.

15 14 13 12 11 10 9 8 7 6 5 4 3 2

Library of Congress Cataloging in Publication Data

Schieler, Leroy.
    Hazardous materials.

    Includes index.
    1. Hazardous substances — Fires and fire prevention.
2. Hazardous substances. I. Pauzé, Denis, joint
author. II. Title.
TH9446.H38S35   604'.7   76.14448
ISBN 0-442-27394-0

PREFACE

The chemist approaches the study of hazardous materials from a frame of reference which explains why the various materials act and react as they do. This explanation is based on the atomic and molecular structure and the chemical reactivity of the materials. Chemists communicate through the use of common nomenclature, chemical symbols, and structural formulas.

The professional fire fighter classifies hazardous materials largely in terms of physical properties and is concerned with ways to transport and store them safely. His greatest challenge is in knowing how to react when the potential danger always present in handling these materials becomes a real danger.

In a progressive industrial nation, the manufacture, processing, handling, storage, and transportation of flammable, explosive, corrosive, and toxic hazardous materials is a way of life. In this situation, it becomes necessary for the fire science technician to integrate basic chemistry concepts with practical fire control methods in order to reach the most successful response to the problem.

HAZARDOUS MATERIALS presents this integrated approach. The reader is introduced to the laws and principles governing the behavior of hazardous materials as a background for learning to control the behavior. Frequently-encountered materials which have hazardous properties are identified, both chemically and practically; rationale for fire prevention and fire fighting is based on both their chemical reactivity and their physical properties. Nationally-accepted procedures for identifying hazardous chemicals and methods for the crisis-handling of them are summarized.

CONTENTS

SECTION 1 BASIC SCIENCES

Unit 1 Atomic Structure and Chemical Reactivity 1
Unit 2 Chemical Formulas and Equations 8
Unit 3 Combustion Mechanisms 11
Unit 4 Endothermic and Exothermic Reactions 14
Unit 5 Heat and the Kinetic-Molecular Theory 17
Unit 6 Gas Laws Governing Temperature, Pressure, and Volume . . . 22
Unit 7 Explosion Mechanisms 25
Unit 8 Shock Waves and Explosions 29
Unit 9 Toxicity 31

SECTION 2 COMBUSTION

Unit 10 Physical Properties Which Affect Combustion Behavior 35
Unit 11 Reactivity With Oxygen 42
Unit 12 Reactivity With Water 45
Unit 13 Reactivity With Chemical Extinguishing Agents 49

SECTION 3 GASES

Unit 14 Compressed Gases 55
Unit 15 Cryogenic Gases 72

SECTION 4 COMBUSTIBLE MATERIALS

Unit 16 Metals 79
Unit 17 Nonmetals 88
Unit 18 Hydrocarbons 93
Unit 19 Substituted Hydrocarbons 100
Unit 20 Plastics 109

SECTION 5 EXPLOSIVE MATERIALS

Unit 21 Principles of Explosives 121
Unit 22 Nitro Explosives 124
Unit 23 Nitric Ester Explosives 127
Unit 24 Other Specialty Explosives 130

SECTION 6 REACTIVE MATERIALS

Unit 25 Peroxides 135
Unit 26 Hydrazines 139
Unit 27 Miscellaneous Reactive Materials 142

SECTION 7 CORROSIVE MATERIALS

Unit 28 Acids 147
Unit 29 Bases 153

SECTION 8 TOXIC MATERIALS

Unit 30 Principles of Toxicology 159
Unit 31 Respiratory Poisons 164
Unit 32 Nerve Poisons 173
Unit 33 Liver Poisons 176
Unit 34 Corrosive Poisons 184

SECTION 9 RADIOACTIVE MATERIALS

Unit 35 Principles of Radioactivity 189
Unit 36 Hazards of Radioactivity 194

SECTION 10 UNIFYING PRINCIPLES

Unit 37 Official Regulations for Handling Hazardous Materials 205
Acknowledgments 245
Index 246

BASIC SCIENCES

Unit 1 Atomic Structure and Chemical Reactivity

Chemistry is the science that deals with matter, reactions, and energy. All matter is composed of atoms. The combination of atoms in various ways to form molecules gives rise to the infinite variety of materials found on earth. To help fire fighters work more safely and effectively with flammable, explosive, toxic, and the other kinds of hazardous materials they often encounter, it is important for them to have a basic understanding of the nature of atoms and of atomic structure. The structure of an atom determines how it combines, or reacts, with other atoms. This reactivity is the prime factor in predicting why certain hazardous materials behave as they do.

THE NUCLEAR ATOM

The smallest and simplest atom is hydrogen (H), which is composed of one proton in the nucleus and one electron orbiting around the nucleus, figure 1-1. Almost all of the mass of any atom is concentrated at the center. This positively-charged body at the center of an atom is called the *nucleus*. The electron orbits around the nucleus just as the earth orbits around the sun. The *electron* which orbits around the nucleus of an atom is identical to an electron which flows through the electrical wiring in a house.

All atoms are electrically neutral, which means that the number of electrons (negative charges) in any atom is always equal to the number of positive charges in the nucleus. The electrons may be thought of as revolving about the nucleus at a high velocity. The *electrostatic attraction,* the force exerted between unlike electrical charges, keeps the negatively-charged electrons from escaping from the positively-charged nucleus.

The *atomic number* of an atom is equal to the number of units of positive charges (protons) in the atomic nucleus. Since the number of electrons is always equal to the number of protons, it can also be said that the atomic number is equal to the number of electrons in the atom.

Thus far, only the simplest of all atoms, the hydrogen atom (composed of one proton and one electron), has been discussed. All other atoms contain a third particle called a *neutron* in addition to protons and electrons. The neutron is a neutral particle in the nucleus of an atom and is very similar to a proton except that it is neutral rather than positively charged.

All atoms of a given element do not necessarily have the same *atomic weight* (total number of protons and neutrons in the nucleus) even though they have the same number

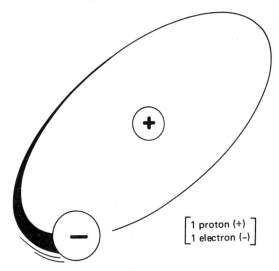

$$\begin{bmatrix} \text{1 proton (+)} \\ \text{1 electron (--)} \end{bmatrix}$$

Fig. 1-1. A Hydrogen Atom.

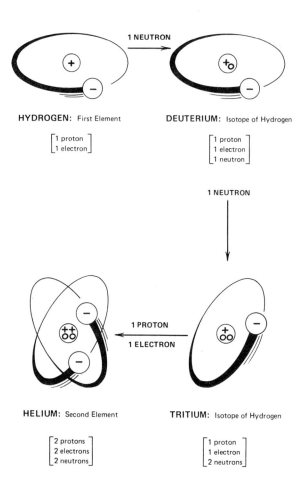

F ig. 1-2. Hydrogen, Its Isotopes, and Helium.

of protons, figure 1-2. This is true because the number of neutrons in the nucleus of a certain element may vary. When this situation is present, the differing atom is called an *isotope* of the element. Although most hydrogen atoms contain only one proton in the nucleus, others called *deuterium* contain one additional neutron, and still others called *tritium* contain two additional neutrons. All are defined as isotopes of hydrogen because they contain one proton and one electron. This similarity in structure makes hydrogen, deuterium, and tritium react alike chemically.

The *elements,* basic atomic building blocks which cannot be broken down into simpler units, are all built up in a similar way from protons, neutrons, and electrons. The first eighteen elements, shown in table 1-1, are the basis for many of the hazardous materials discussed in this text.

ELECTRONIC CONFIGURATION OF ATOMS

The arrangement and behavior of electrons in atoms is very complex. The elec-

Atom		Atomic Number	Mass Number	Nucleus		
Name	Symbol			Protons	Neutrons	Electrons
Hydrogen	H	1	1	1	0	1
Helium	He	2	4	2	2	2
Lithium	Li	3	7	3	4	3
Beryllium	Be	4	9	4	5	4
Boron	B	5	11	5	6	5
Carbon	C	6	12	6	6	6
Nitrogen	N	7	14	7	7	7
Oxygen	O	8	16	8	8	8
Fluorine	F	9	19	9	10	9
Neon	Ne	10	20	10	10	10
Sodium	Na	11	23	11	12	11
Magnesium	Mg	12	24	12	12	12
Aluminum	Al	13	27	13	14	13
Silicon	Si	14	28	14	14	14
Phosphorus	P	15	31	15	16	15
Sulfur	S	16	32	16	16	16
Chlorine	Cl	17	35	17	18	17
Argon	Ar	18	40	18	22	18

Table 1-1. Structure of the First Eighteen Elements.

Atom		Atomic Number	Electrons		
Name	Symbol		1st Layer	2nd Layer	3rd Layer
Hydrogen	H	1	1		
Helium	He	2	2		
Lithium	Li	3	2	1	
Beryllium	Be	4	2	2	
Boron	B	5	2	3	
Carbon	C	6	2	4	
Nitrogen	N	7	2	5	
Oxygen	O	8	2	6	
Fluorine	F	9	2	7	
Neon	Ne	10	2	8	
Sodium	Na	11	2	8	1
Magnesium	Mg	12	2	8	2
Aluminum	Al	13	2	8	3
Silicon	Si	14	2	8	4
Phosphorus	P	15	2	8	5
Sulfur	S	16	2	8	6
Chlorine	Cl	17	2	8	7
Argon	Ar	18	2	8	8

Table 1-2. Electronic Configuration of the First Eighteen Elements.

trons arrange themselves around the dense, positively-charged nucleus in the pattern shown in table 1-2.

It was mentioned earlier that the number of electrons orbiting around the nucleus of an atom is equal to the number of protons in the nucleus and that the atomic number of an atom is the same as the number of protons (or electrons) in the atom. When the electrons of an atom arrange themselves around the nucleus according to the pattern just established, some-

times the outer layer is full, table 1-3. More often, it is not, as can be seen in table 1-2.

LAWS OF CHEMICAL CHANGE

Compounds are substances containing two or more elements that are chemically combined. For example, the compound *methane* (marsh gas) contains the elements carbon and hydrogen; the compound *sodium chloride* (salt) contains the elements sodium and chlorine; the compound *sucrose* (sugar)

Atomic Number	Element		1st Layer	2nd Layer	3rd Layer	4th Layer	5th Layer	6th Layer
	Symbol	Name						
2	He	Helium	2					
10	Ne	Neon	2	8				
18	Ar	Argon	2	8	8			
36	Kr	Krypton	2	8	18	8		
54	Xe	Xenon	2	8	18	18	8	
86	Rn	Radon	2	8	18	32	18	8

Table 1-3. Electronic Configuration of the Inert Gas Family of Elements.

contains the elements carbon, hydrogen, and oxygen.

Compounds can be decomposed into the elements from which they are chemically formed. In contrast, elements cannot be decomposed into smaller units by chemical methods. Thus, elements are considered the building blocks of compounds.

Mixtures are made up of constituents present in varying proportions. If salt and sugar are mixed, no changes are produced in the individual constituents. No matter how finely they are powdered, they are still salt and sugar. The properties of mixtures are variable whereas those of compounds are constant.

Physical changes are changes such as melting, boiling, or freezing which do not change the chemical composition of the material. Ice can be melted to water, and water can be boiled to steam, but ice, water, and steam all retain the same chemical composition.

Magnesium + Oxygen → Magnesium oxide

(element) + (element) → (compound)

A closer study of chemical reactions such as the formation of magnesium oxide from magnesium and oxygen reveals definite fundamental laws. The first law, *the law of conservation of mass,* states that matter is neither created nor destroyed in chemical reactions but merely changes into different forms. The second, *the law of constant composition*, states that chemical reactants combine in definite weight ratios.

CHEMICAL REACTIONS

In chemical reactions, electrons of the elements are shared so that the compound formed has a *closed shell structure* (the outer layer is completed).

- H shares electrons so that it has two outer electrons like He:.

- All other atoms share electrons so that they have eight outer electrons like

:Ne: :Ar: :Kr: :Xe:

An example of a common reaction will help to make the meaning clearer.

H· + H· → H:H

This is the reason that chemists write hydrogen as H_2 rather than H. Hydrogen atoms combine to form the *diatomic molecule* (a molecule formed from two atoms) H_2 (H·H). The hydrogen found in a gas bottle is H_2, not H, a fact that has important consequences in combustion and explosion mechanisms.

All chemical reactions occur in the same way, with hydrogen sharing two electrons and all other elements sharing eight electrons. All of the common gases (oxygen, nitrogen, fluorine, and chlorine) form diatomic molecules by sharing eight electrons (an *octet*) in a manner similar to that shown for hydrogen.

O: + O: → O::O

N: + N: → N:::N

·F: + ·F: → :F:F:

·Cl: + ·Cl: → :Cl:Cl:

Chemical reactions involving atoms of different elements occur by application of the same electron-sharing principles. Carbon and oxygen rearrange to form carbon dioxide in a combustion reaction. Carbon monoxide is formed from the same elements when proportionately less oxygen is present.

C + O_2 → CO_2

Carbon + Oxygen → Carbon dioxide

C: + O::O → O::C::O

$$2C \quad + \quad O_2 \quad \rightarrow \quad 2CO$$

Carbon + Oxygen → Carbon Monoxide

$$2 \overset{\cdot\cdot}{C}{:} \quad + \quad \overset{\cdot\cdot}{\underset{\cdot\cdot}{O}}{::}\overset{\cdot\cdot}{\underset{\cdot\cdot}{O}} \quad \rightarrow \quad 2 : C:::O:$$

THE PERIODIC LAW

Since over 100 different elements are known, the task of learning their properties and those of their compounds would be almost impossibly difficult without a method of classification into smaller groups of similar elements. Fortunately, such a simplification is available through the *Periodic Law* which states that the physical and chemical properties of the elements are periodic functions of their atomic numbers. A periodic chart (arranged so that all elements which appear in a given group contain the same number of electrons in the outer orbit and thus have similar chemical and physical properties) is used to illustrate this law. Figure 1-3 shows the chart with the elements arranged into given groups, such as metals, nonmetals, and gases, which makes it possible to estimate information such as atomic number, boiling point, melting point, density, and stability.

In the case of chemical properties, the periodic relationship applies when the elements are listed according to their atomic number sequence. The elements which occupy first place in each of the several periods indicated on the chart are lithium, sodium, potassium, rubidium, cesium, francium. These elements are not only similar in physical properties, but are all very reactive metals which exhibit similar chemical reactivity. The elements which occupy last place in each period are helium, neon, argon, krypton, xenon, and radon. These elements are all nonreactive gases which exhibit a similar degree of *chemical inertness* (do not combine with other elements).

Group I	II											III	IV	V	VI	VII	VIII
1 H																	2 He
3 Li	4 Be											5 B	6 C	7 N	8 O	9 F	10 Ne
11 Na	12 Mg											13 Al	14 Si	15 P	16 S	17 Cl	18 Ar
19 K	20 Ca	21 Sc	22 Ti	23 V	24 Cr	25 Mn	26 Fe	27 Co	28 Ni	29 Cu	30 Zn	31 Ga	32 Ge	33 As	34 Se	35 Br	36 Kr
37 Rb	38 Sr	39 Y	40 Zr	41 Nb	42 Mo	43 Tc	44 Ru	45 Rh	46 Pd	47 Ag	48 Cd	49 In	50 Sn	51 Sb	52 Te	53 I	54 Xe
55 Cs	56 Ba	57-71 La*	72 Hf	73 Ta	74 W	75 Re	76 Os	77 Ir	78 Pt	79 Au	80 Hg	81 Tl	82 Pb	83 Bi	84 Po	85 At	86 Rn
87 Fr	88 Ra	89-103 Ac*															

Alkali Metals (left margin label)

Inert Gases (right margin label)

Fig. 1-3. A Simple Example of the Periodic Chart of the Elements.

La* — a series of rare earth elements which all react the same chemically.

Ac* — a series of radioactive elements which all react the same chemically.

Atomic Number	Element		Electronic Configuration						
	Symbol	Name	1st Layer	2nd Layer	3rd Layer	4th Layer	5th Layer	6th Layer	7th Layer
3	Li	Lithium	2	1					
11	Na	Sodium	2	8	1				
19	K	Potassium	2	8	8	1			
37	Rb	Rubidium	2	8	18	8	1		
55	Cs	Cesium	2	8	18	18	8	1	
87	Fr	Francium	2	8	18	32	18	8	1

Table 1-4. Electronic Configuration of the Alkali Metal Family of Elements.

A resemblance is evident among any group of elements (or compounds of the elements) which occupy similar positions in the respective periods, table 1-4. This similarity indicates that it is not the total number of electrons in an atom that determines the chemical reactivity of the elements, but the number in the outer layer.

Generalizations of this type are valuable to the study of hazardous materials in that they provide a basis for understanding the properties of known materials and predicting the behavior of less well known ones.

The periodic table is used as a guide by the chemist to predict the chemical reactivity and physical characteristics of any element. It is analogous to the hazardous materials handbook that the fire fighter uses. The integration of both of these provides usable guidelines for handling hazardous materials.

REVIEW

1. All atoms are

   a. Electrically positive.
   b. Electrically negative.
   c. Electrically neutral.
   d. A combination of positive, negative, and neutral.

2. The atomic number of an atom is equal to

   a. The number of electrons and neutrons.
   b. The number of protons.
   c. The number of neutrons.
   d. The sum of the number of electrons, protons, and neutrons.

3. An isotope of sodium would contain

   a. More electrons.
   b. More protons.
   c. More neutrons.
   d. More neutrons and protons.

4. The elements are composed of

    a. Protons, neutrons, and electrons.
    b. Neutrons and electrons.
    c. Hydrogen atoms.
    d. Hydrogen isotopes.

5. An element can be chemically broken down by

    a. Fire.   c. Acids.
    b. Explosions.   d. None of these.

6. Which of the following is a chemical change?

    a. Water → Steam.
    b. Water → Ice.
    c. Sodium and oxygen → sodium oxide.
    d. All of the above.

7. As shown in the periodic chart (figure 1-3), which of the following groups of elements would have similar chemical properties?

    a. Li, Na, and Ca.   c. O, F, and Cl.
    b. Be, Mg, and Ca.   d. He, Ne, and N.

8. As shown in the periodic table (figure 1-3), which of the following pairs of elements would have similar chemical properties?

    a. Be and Mg.   c. Li and Na.
    b. F and Cl.   d. All of these.

Unit 2 Chemical Formulas and Equations

Molecules are formed by the combination of atoms. In the same way that symbols are used to represent atoms, atomic symbols are written together to represent compounds. These groupings of symbols are referred to as *chemical formulas* or *molecular formulas*.

SYMBOLS FOR ATOMS		MOLECULAR FORMULAS	
Hydrogen	H	Water	H_2O or H-O-H
Sodium	Na	Sodium chloride (salt)	NaCl or Na-Cl
Beryllium	Be	Ammonia	NH_3 or H-N-H with H below
Chlorine	Cl		
Oxygen	O	Beryllium oxide	BeO or Be-O
Nitrogen	N		

The two types of molecular formulas shown represent the same thing. The formula for water indicates that it is made up of two atoms of hydrogen and one atom of oxygen. Similarly, salt is composed of one atom of sodium and one of chlorine.

COMBINING CAPACITY OF ATOMS

Atoms of elements combine to form molecules of compounds. The combination (reactions) takes place in predictable ratios. The number which represents the chemical combining capacity of a given element is its *valence*. The combining capacity is related to the electronic configuration of the atom, with emphasis on the outer layer of electrons.

HCl	CaO	$FeCl_3$
NaCl	$CaCl_2$	Fe_2O_3
H_2O	CO_2	$AlCl_3$
Na_2O	CCl_4	Al_2O_3

The simplifying assumption is made that the valence of hydrogen (H) is 1. With this assumption and the known molecular formula of the compounds, the valence of all of the elements can be deduced.

The valence of chlorine (Cl) is 1 because one atom is combined with 1 hydrogen (H) atom.
The valence of sodium (Na) is 1 because one atom is combined with 1 chlorine (Cl) atom.
The valence of oxygen (O) is 2 because one atom is combined with 2 hydrogen (H) atoms.
The valence of calcium (Ca) is 2 because one atom is combined with 1 oxygen (O) atom.
The valence of carbon (C) is 4 because one atom is combined with 2 oxygen (O) atoms.
The valence of both iron (Fe) and aluminum (Al) is 3 because, in each case, one atom is combined with 3 chlorine (Cl) atoms.

Some elements can combine in more than one weight ratio. A number of elements have more than one valence. Some have 4 or more. Perhaps the most familiar examples are carbon dioxide and carbon monoxide.

CO_2	carbon with a valence of 4
CO	carbon with a valence of 2

NAMING COMPOUNDS

It is necessary to establish a preliminary convention for naming chemical compounds in order to establish a vocabulary. The names of *binary compounds*, compounds containing two elements, usually end in "ide."

HCl	Hydrogen chlor<u>ide</u>
NaCl	Sodium chlor<u>ide</u> (salt)
H_2O	Hydrogen ox<u>ide</u> (water)
Na_2O	Sodium ox<u>ide</u>
CaO	Calcium ox<u>ide</u>
$CaCl_2$	Calcium chlor<u>ide</u>
CO_2	Carbon diox<u>ide</u>

If an element has more than one valence, the lower valence is indicated by the suffix "ous" and the higher valence by the suffix "ic."

$FeCl_2$	ferr<u>ous</u> chloride
$FeCl_3$	ferr<u>ic</u> chloride
$HgCl$	mercur<u>ous</u> chloride
$HgCl_2$	mercur<u>ic</u> chloride

CHEMICAL EQUATIONS

Chemical equations are a shorthand, or convenient, way of describing chemical reactions. The symbols for the elements and the formulas for the compounds are the shorthand notation. The reaction of hydrogen and oxygen to form water, in chemical notation, is expressed:

$$H_2 + O_2 \rightarrow H_2O$$

This chemical notation is incomplete in the sense that the number of H atoms and O atoms in the reactants are not equal to those in the products. There are two H atoms and two O atoms in the left side of the equation, but only one O atom appears on the right side. The equation must be *balanced*. In a balanced chemical equation, the numbers of atoms of reactants and products must be equal. This was discussed previously and was called the Law of Conservation of Mass. A properly balanced equation shows this equality.

$2H_2 + O_2 \rightarrow 2H_2O$		coefficient
4 hydrogen atoms	4 hydrogen atoms	$2H_2$
2 oxygen atoms	2 oxygen atoms	subscript

A chemical equation is balanced by placing the proper *coefficient* in front of the formula. This is done by trial and error. An important point to remember is that the subscript numbers in the formula must never be changed in attempting to balance a chemical equation. The formula represents the composition of the compound and is unchanging. Placing a coefficient in front of the formula changes the number of units of the compound entering into the reaction, but it does not change the composition of the unit. The process of balancing equations seems simpler after studying a few examples and practicing the process.

EXAMPLE 1

$Na + H_2O \rightarrow NaOH + H_2$
Step 1: $Na + 2H_2O \rightarrow NaOH + H_2$
Step 2: $Na + 2H_2O \rightarrow 2NaOH + H_2$
Step 3: $2Na + 2H_2O \rightarrow 2NaOH + H_2$

This is an important equation in the study of hazardous materials. The balanced equation indicates that 2 atoms of sodium (Na) react with 2 molecules of water (H_2O) to form 2 molecules of sodium hydroxide (NaOH) and 1 molecule of hydrogen (H_2). This is why the National Fire Protection Association uses a W̶ in the information triangle. This indicates that water should never be used in the presence of sodium because the resulting reaction creates hazardous materials: sodium hydroxide (commonly known as lye), which is a caustic, and hydrogen, which is explosive and flammable. This reaction has caused many fires in chemical and industrial laboratories.

EXAMPLE 2

$H_2 + F_2 \rightarrow HF$
Step 1: $H_2 + F_2 \rightarrow 2HF$

This equation indicates that 1 molecule of hydrogen reacts with 1 molecule of fluorine to form 2 molecules of hydrogen fluoride.

9

REVIEW

1. The valence of carbon in methane (CH_4) is

   a. 2.   c. 1.
   b. 4.   d. 5.

2. The valence of hydrogen in ethane ($H_3C\text{-}CH_3$) is

   a. 2.   c. 6.
   b. 4.   d. 1.

3. Which of the following chemical equations is *not balanced*?

   a. $HCl + NaOH \rightarrow NaCl + H_2O$ c. $O_2 + 2Na \rightarrow Na_2O$
   b. $2HCl + 2Na \rightarrow 2NaCl + H_2$ d. $2H_2O + 2Na \rightarrow 2NaOH + H_2$

4. Sodium reacts violently with

   a. Magnesium.   c. Hydrogen.
   b. Water.   d. Titanium.

5. How many atoms of oxygen are there in a molecule of carbon monoxide?

   a. 1   c. 3
   b. 2   d. 4

6. A chemical equation is balanced by

   a. Trigonometry.   c. Neutrons.
   b. Protons.   d. Trial and error.

7. The formula for the oxygen molecule is

   a. O.   c. O_3.
   b. O_2.   d. Cl_2.

8. The formula for a chlorine atom is

   a. Cl.   c. NaCl.
   b. Cl_2.   d. KCl.

9. Elements can combine

   a. Only in the same weight ratio.   c. In more than one weight ratio.
   b. With only the same elements.   d. Only half their weight.

10. Carbon dioxide and carbon monoxide have

    a. The same weight.   c. Different valences for carbon.
    b. The same valence for carbon.   d. Equal densities.

11. Chemical equations

    a. Describe chemical reactions.   c. Are always incomplete.
    b. Contain only two elements.   d. All of these.

12. The molecular formula for water is

    a. H_2O.   c. 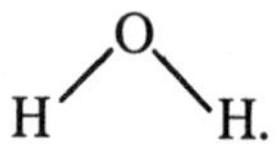 H$\diagup$O$\diagdown$H.
    b. H-O-H.   d. All of these.

Unit 3 Combustion Mechanisms

In the fire sciences, combustion processes are usually explained in terms of the fire triangle or the fire rectangle, figure 3-1. In the fire triangle representation, combustion is pictured as the combination of oxygen and fuel with the release of heat. The combustion of methane in oxygen with the formation of carbon dioxide, water, and heat may be shown by the chemical equation or by the fire triangle.

$$CH_4 + 2O_2 \rightarrow CO_2 + 2H_2O + heat$$

fuel   oxygen

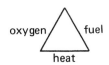

This seems obvious enough and is correct as far as it goes! However, the fire triangle concept (or the chemical equation because it represents the same thing) fails to explain several generally accepted facts in the area of fire control.

- That dry chemical extinguishers are effective.

  a. Dry chemical extinguishers do not seem to extinguish a fire by cooling it because the relative efficiency of various types is not correlated to their heat capacities. (Stated another way, they do not seem to function like a heat sink and, thereby, remove heat from the flame.)

  b. Dry chemical extinguishers do not seem to function by liberating carbon dioxide and water in the flame and thus keep oxygen away from the flame. (Stated another way again, they do not seem to exclude oxygen from the flame and thus suffocate it.)

  Note: Dry chemical extinguishers are usually sodium bicarbonate ($NaHCO_3$), potassium bicarbonate ($KHCO_3$), or monoammonium phosphate. It was thought for a long time that the first two functioned by breaking down in the fire with the formation of carbon dioxide and water, thus smothering the fire. This was shown not to be true by demonstrating that the amount of carbon dioxide and water generated by a given weight of dry chemical extinguisher was much less effective.

  c. It does not explain why potassium bicarbonate is much more efficient than sodium bicarbonate as a dry chemical extinguisher.

- That halogenated extinguishing agents (Halon series) are good extinguishers.

- That a fire burns in the first place!

The fire triangle concept is not wrong; it simply does not provide enough detail about the overall combustion mechanism. In order to understand combustion and how dry chemical and halogenic extinguishers work, as well

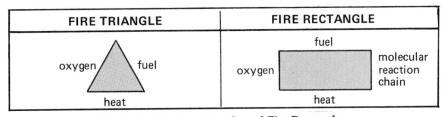

Fig. 3-1. The Fire Triangle and Fire Rectangle.

as to benefit from much of the fire science material to follow, it is necessary to consider the combustion concept in greater detail.

In the combustion of methane (natural gas) with oxygen, the overall course of the combustion reaction is represented by the chemical equation:

$$CH_4 + 2O_2 \rightarrow CO_2 + 2H_2O + heat$$

The mechanism of the combustion reaction involves a number of intermediate steps, however. The methane and oxygen are broken down in successive steps until they finally arrive at their end point as carbon dioxide and water:

$$CH_4 \overset{heat}{\rightarrow} CH_3^{\cdot} + H^{\cdot} \rightarrow CH_2^{\cdot} + H^{\cdot} \rightarrow$$
$$CH^{\cdot} + H^{\cdot} \rightarrow C + H^{\cdot}$$
$$O_2 \overset{heat}{\rightarrow} O^{\cdot} + O^{\cdot}$$

These partially broken down molecules are called *free radicals:* $H^{\cdot}$ is called a hydrogen free radical; O, an oxygen free radical; and C, a carbon free radical. The point to consider is that they are partially broken down molecules which are much more reactive chemically than the methane or the oxygen.

These free radicals, or intermediate combustion products, react with each other and with CH_4 and O_2 until they are ultimately converted to CO_2 and water, releasing heat in the process. Without trying to show every step as a balanced chemical reaction, for the sake of simplicity, the sequence of free radical reactions may be represented as follows:

$$CH_4 + O^{\cdot} \rightarrowtail CO_2 + H_2O + heat$$
$$CH_3^{\cdot} + O^{\cdot} \rightarrowtail CO_2 + H_2O + heat$$
$$CH_2^{\cdot} + O^{\cdot} \rightarrowtail CO_2 + H_2O + heat$$
$$CH^{\cdot} + O^{\cdot} \rightarrowtail CO_2 + H_2O + heat$$
$$C + O^{\cdot} \rightarrowtail CO_2 + heat$$
$$H^{\cdot} + O^{\cdot} \rightarrowtail H_2O + heat$$
$$H_2 + O^{\cdot} \rightarrowtail H_2O + heat$$
$$H^{\cdot} + O_2 \rightarrowtail H_2O + heat$$

It should be noted that methane (CH_4) and oxygen (O_2) are not shown as reacting directly with each other. This reaction is slow compared to the free radical reactions of the type shown. Actually then, combustion proceeds via a complex series of free radical reactions — not by a direct reaction.

This same concept was pictured in non-chemical terms by the more familiar fire rectangle shown in figure 3-1. It may be instructive to note that while the fire fighter may never have thought of a fire in terms of free radicals, the chemist probably has never thought of it in terms of fire rectangles.

Scientists have proved that the free radical combustion mechanism is correct. This conceptual framework helps explain some of the commonly used techniques for controlling fires.

- Why a hydrocarbon like gasoline can be mixed with air, which is 20-percent oxygen, and combustion does not start until a spark or other initiator is provided: Gasoline and oxygen do not react directly. They react via free radicals. The spark or initiating source partially breaks down a few molecules, providing the free radicals which initiate the combustion reaction. The heat generated at each step of the reaction creates more free radicals which propagate the combustion. The same logic applies to any chemical combustion or explosion reaction.

- Why a fire can be put out by excluding oxygen with water or sand: The formation of enough free radicals to propagate the combustion is prevented.

- Why dry chemical extinguishers are so effective: The sodium and potassium portions of the bicarbonates are free radical quenchants. They help stop the fire by reducing the concentration of free radicals – as well as by excluding oxygen.

- Why potassium is better than sodium bicarbonate as a dry chemical extinguisher: Potassium is a better free radical quenchant than sodium. (It will be shown later why this is true.)
- Why Halon® is an efficient extinguisher: The halogens – chlorine, bromine, fluorine, and iodine – are among the most efficient free radical quenchants. The Halon series, such as Halon 1301, employs the use of bromine as quenchants.

Doubtless, the student will have many other questions – if not now, certainly later. However, he may sense at this relatively early point, that a fundamental background in chemical reactivity can give a hitherto unsuspected insight into fire protection techniques. He will find it to be a basis for the continued study of hazardous materials.

REVIEW

1. Which of the following are free radicals?
   a. $CH_3 \cdot$
   b. $CH_2 \cdot$
   c. $CH \cdot$
   d. All of these

2. The formula $CH_3 \cdot$ represents a
   a. Hydrogen free radical.
   b. Carbon free radical.
   c. Methyl free radical.
   d. None of these.

3. A mixture of gasoline and air does not ignite until a spark is applied because
   a. An initiating source is required to generate free radicals.
   b. Gasoline and oxygen do not react spontaneously.
   c. The temperature is too low to create enough free radicals to initiate the reaction.
   d. All of the above.

4. Dry chemical fire extinguishers are very effective because
   a. They exclude oxygen from the combustion zone.
   b. The sodium portion of sodium bicarbonate is an effective quenchant.
   c. The bicarbonate portion of sodium bicarbonate is an effective quenchant.
   d. The bicarbonate portion of sodium bicarbonate breaks down to liberate CO_2.

5. The formula for sodium bicarbonate is
   a. $NaCO_3$.
   b. Na_2CO_3.
   c. $NaHCO_3$.
   d. $Na(CO_3)_2$.

6. The best way to explain the efficiency of the halogen fire extinguishers is in terms of
   a. The fire triangle.
   b. The fire rectangle.
   c. The low carbon content of halogen extinguishers.
   d. The low chemical reactivity of halogens with O_2.

Unit 4 Endothermic and Exothermic Reactions

Fire fighters should be aware of the two general types of reactions that occur: exothermic and endothermic. A process or chemical reaction which liberates heat is called an *exothermic* reaction. This type of reaction is encountered daily by the fire fighter and any number of examples may be used, such as gasoline or propane fires. In these fires, the intense heat given off provides evidence that an exothermic reaction is happening. Conversely, processes or reactions which absorb heat are called *endothermic* reactions. Here, heat must be supplied to the reaction to make it occur. The cold pack carried in rescue vehicles and ambulances works on the similar principle of an endothermic reaction. When the packet is squeezed together and the contents mixed, heat is absorbed as the chemical process takes place. The pack becomes cold.

Fire suppression depends heavily on endothermic processes. For example, when water is sprayed on a fire, it is cooled because the water is changed to vapor. This vaporization process requires a great amount of heat.

$$H_2O \text{ (liquid)} \text{ and Heat} \rightarrow H_2O \text{ (gas)}$$

As the heat is withdrawn from the combustion zone, the temperature of the material is brought below its ignition temperature and flash point, and the fire is suppressed.

Water does not always favor endothermic processes; at times, it enters into exothermic reactions. If water is sprayed on an extremely hot fire, such as a magnesium fire, the temperature is great enough to dissociate the water. The end result is a rapid, violent exothermic reaction. In this reaction, magnesium hydroxide and hydrogen are formed. Fuel (H_2) and oxidizer (O_2) are added to the fire when water is applied.

$$Mg + 2H_2O \rightarrow Mg(OH)_2 + H_2$$

Water is an excellent extinguishing agent in actual practice, however, since most fires are not hot enough or chemically active enough to dissociate the water. Generally, the water vaporizes, locally cooling the flame, and expands away from the fire zone before it can be dissociated.

Other extinguishing agents, such as the dry chemicals which have already been said to exclude oxygen and quench free radicals, can be seen as participating in endothermic reactions. Sodium bicarbonate ($NaHCO_3$), for example, removes heat from the reaction zone, and therefore the reaction is endothermic. The heat dissociates the sodium bicarbonate into more effective extinguishing agents — water and carbon dioxide.

$$2NaHCO_3 + Heat \rightarrow Na_2CO_3 + CO_2 + H_2O$$

Exothermic reactions give off heat. But why do some substances give off more heat than others? The amount of heat — measured

Exothermic Reaction	Heat Liberated (kcal)	Endothermic Reaction	Heat Absorbed (kcal)
$2H_2 + O_2 \rightarrow 2H_2O + heat$	116	$N_2 + O_2 + heat \rightarrow 2NO$	43
$C + O_2 \rightarrow CO_2 + heat$	94		
$2CO + O_2 \rightarrow 2CO_2 + heat$	136	$CO_2 + heat \rightarrow C + O_2$	94
$CH_4 + 2O_2 \rightarrow CO_2 + 2H_2O + heat$	213		
$2Mg + O_2 \rightarrow 2MgO + heat$	180	$2H_2O + heat \rightarrow 2H_2 + O_2$	116

Fig. 4-1. Examples of Some Exothermic and Endothermic Reactions.

in kilocalories (kcal) — produced during combustion is directly related to the *bond energies* of the material. These bonds, or "links," which hold the molecules of compounds together, are of varying strength for each particular molecule and, therefore, have different energy values assigned to them, table 4-1, page 14.

Hydrogen (H_2), for example, has a bond energy of 104 kcal. It is a very stable molecule because it is held together very strongly. Hydrogen has a relatively high bond energy when compared to the other bond strengths listed in table 4-2. When it does enter into a chemical reaction, it has a lot of energy to release and tends to be exothermic.

The Houston train derailment described here is an example of what can happen when energy bonds are broken, rearranged, and their energy released.[1]

On October 19, 1971, a freight train derailed inside the city limits of Houston, Texas. Among the cars that derailed were six tank cars of vinyl chloride, three of fuel oil,

H–H	104	C–H	99	H–O	111
C–C	83	C–O	84	H–N	94
N–N	38	C–N	70	H–F	135
O–O	34	C–F	105	H–Cl	103
F–F	38	C–Cl	78	H–S	88
Cl–Cl	58				
S–S	64				

Table 4-2. Bond Energies in kcal Per Mole.

and one each of acetone, butadiene, and formaldehyde. Fire from leaking vinyl chloride heated another tank car containing 48,000 gallons of vinyl chloride causing heat and pressure buildup within the tank car. The sudden release of this energy, despite efforts to cool the tank car, resulted in the car rupturing with a fireball that was reported to be 1,000 feet in diameter. The heat potential was estimated to be some 800 million BTU[1]. Before the incident was over, one training officer was killed, fourteen fire fighters were injured, and one pumper was destroyed.

REVIEW

1. In the chemical reaction $2H_2 + O_2 \rightarrow 2H_2O + heat$
   a. Energy is released.
   b. New products are formed.
   c. Reactants are consumed.
   d. All of these.

2. In the chemical reaction $N_2 + O_2 + heat \rightarrow 2NO$
   a. Energy is consumed.
   b. New products are formed.
   c. Reactants are consumed.
   d. All of these.

3. Which of the following reactions is exothermic?
   a. $2H_2O + heat \rightarrow 2H_2 + O_2$
   b. $H_2 + heat \rightarrow 2H$
   c. $2H \rightarrow H_2 + heat$
   d. $CO_2 + heat \rightarrow C + O_2$

4. Which of the following classes of chemical reactions are usually exothermic?
   a. Combustion
   b. Explosion
   c. Corrosion
   d. All of these

[1]Hayes, Terry C., "Houston Train Derailment and the Subsequent Hearing," *Fire Journal* Vol. 66. No. 2 (March 1972) p. 17.

5. Under some conditions water can intensify a fire because

    a. The water was not precooled enough.
    b. Too much water was used.
    c. The fire dissociated the water.
    d. Too little water was used.

6. . A compound with a high bond energy usually

    a. Releases much energy when burned.
    b. Releases little energy when burned.
    c. Is too stable to burn.
    d. Is endothermic.

7. Which of the following has the lowest bond energy?

    a. F_2.   c. H_2.
    b. O_2.   d. Cl_2.

8. Which of the following has the highest bond energy?

    a. H_2.   c. HCl.
    b. HF.   d. F_2.

9. The bond energy of a molecule is most closely related to its

    a. Exothermic reactions.   c. Valence.
    b. Endothermic reactions.   d. Bond strength.

Unit 5 Heat and the Kinetic - Molecular Theory

The chemical theories which were developed in the preceding units are valuable even though each molecule was characterized only in terms of its atomic structure. In order to take the next step in the study of hazardous materials, it is necessary to learn more about the nature of molecules, particularly gas molecules. In the same sense that chemical reactivity is best discussed within the framework of atomic structure, heat is best discussed within the framework of the kinetic-molecular theory. It is this theory that explains the properties of natural gas (methane) and its differences from propane and butane.

KINETIC-MOLECULAR THEORY

All gases are made up of atoms and molecules. Simplified pictures of an atom and a molecule were developed in preceding units. Both atoms and molecules are very small; molecules of any given gas are all the same mass and size. For example, all oxygen molecules (O_2) are the same mass and size; all hydrogen molecules (H_2) are the same size. However, oxygen molecules are heavier and larger than hydrogen molecules as shown in figure 5-1.

The space occupied by the molecules of a gas is a small percentage of its total volume. All the molecules are in constant random motion; in other words, they have no direct path. They collide with each other and with the walls of a container in random fashion. Between collisions they always move in straight line paths. The collisions of all atoms and molecules are perfectly elastic. The simplest way of looking at them is to compare them to the collision of two billiard balls, figure 5-2.

Pressure

Many of the phenomena dealt with in fire science — pressure, boiling, evaporation, cooling, melting, and freezing — are more easily understood in terms of the kinetic-molecular theory. Pressure may be used as an example. When a balloon or an automobile tire is inflated, what is happening? According to the kinetic-molecular theory, the pressure exerted by a gas is due to the collision of gas molecules against the wall of the container. It is the same principle as that of the pressure exerted by raindrops falling on a roof or of the wind (air molecules) colliding with a door. The motion of the atoms or molecules of a gas is random; therefore the pressure is uniform in all directions. This is the reason a balloon or

hydrogen oxygen water

$$2H_2 \;+\; O_2 \;\rightarrow\; 2H_2O$$

Fig. 5-1. Relative Size of Some Common Atoms and Molecules.

BEFORE COLLISION	AFTER COLLISION

Fig. 5-2. Elastic Collision of Atoms and Molecules.

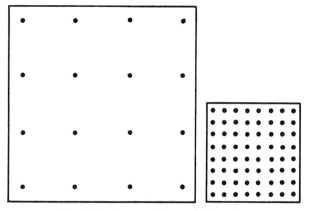

Fig. 5-3. Pressure Exerted By a Gas.

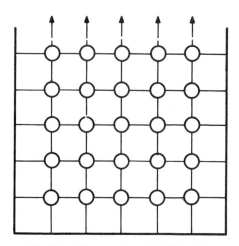

Fig. 5-4. Evaporation of a Liquid.

tire can be inflated. Since there is a large amount of unoccupied space between the atoms or molecules of a gas, the gas is highly compressible. This principle is illustrated in figure 5-3. When the volume of a fixed volume of gas is reduced by half, the pressure is doubled because twice as many collisions per unit area occur.

Physical States

The kinetic-molecular theory is not limited to gases, but is equally applicable to liquids and solids. If a gas is cooled to progressively lower temperatures, it will first liquefy and then solidify. This again is a direct consequence of kinetic-molecular theory. As a gas cools, its kinetic motion decreases. When kinetic motion is sufficiently lowered, the molecules get close enough that intermolecular forces hold them together to form the state of matter called a liquid. Further cooling of the liquid brings the molecules still closer together and the liquid solidifies. These changes take place in reverse when heat is added. Ice melts to form water; water boils to form steam.

Evaporation and Boiling of Liquids

When water is placed in an open dish, it gradually decreases in volume. It evaporates. A rapidly moving molecule near the surface of the liquid may have sufficient kinetic energy

to overcome the attraction of its neighboring molecules and escape. The rate of evaporation increases with increasing temperatures because molecular motion increases with temperature. As shown in figure 5-4, molecules which reach the surface are restricted on only three sides, but in the interior of the liquid, they are confined on all sides.

When the space above the liquid is confined, as in a closed bottle, molecules cannot completely escape but strike the walls of the container and condense. The rate of condensation soon becomes equal to the rate of evaporation and a liquid-vapor equilibrium is established.

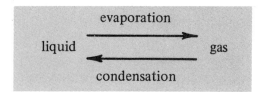

At any given temperature, the pressure exerted by the liquid-gas equilibrium is called the *vapor pressure.*

As temperature is increased, a point is reached at which vapor pressure is equal to the surrounding atmospheric pressure. For butane, this is 32°F. At sea level, water boils at 212°F where the pressure is one atmosphere. On a mountain top where the pressure is lower, the boiling point is lower. Food cooked in boiling water at higher altitudes cooks more slowly because the temperature of the boiling water is less than at sea level. Similarly, the boiling point of water is higher at higher pressures as in a pressure cooker.

Melting of Solids

When a solid is heated, the molecular motion is increased sufficiently to overcome the attractive forces holding the solids together, and melting occurs. If heating is continued, all of the solid passes into the liquid state. However, if heating is stopped before melting is complete, the liquid and solid phases remain in equilibrium. The thawing of ice is an example. The temperature at which the liquid and solid phases are in equilibrium is called the freezing point of the liquid or the melting point of the solid. Numerically, they are the same.

Solids have vapor pressures just as liquids do. Again, this is a necessary consequence of the kinetic-molecular theory. Moth balls (paradichlorobenzene) pass directly from the solid state to the gaseous state. This is called *sublimation*. Snow and ice sublime or pass directly from the solid state to the gaseous state. This is the reason snow and ice evaporate below their melting point. Another substance, dealt with extensively in the fire

Rate	Type of Chemical Reaction
Almost instantaneous	Detonations
Very rapid	Acid-base neutralizations
Rapid	Explosions
Average	Combustion
Slow	Corrosion of metals

Fig. 5-5. Rates of Typical Chemical Reactions Encountered in the Fire Sciences.

sciences, which sublimes is dry ice (carbon dioxide).

Chemical Reactivity

The speed with which a chemical reaction takes place is directly related to the kinetic motion of the molecules involved. It was said earlier that an increase in pressure brings the molecules closer together and so increases the chances of collision. Likewise, an increase in temperature by increasing the speed of the motion makes collision more certain. A third condition, an increase in the concentration of reactants, also causes more frequent collisions of molecules. Any increased molecular collision rate causes a chemical reaction to proceed more rapidly, figure 5-5. What is the significance of this to the person handling hazardous materials? It helps in predicting and preparing for situations in which reactive materials are exposed to heat and increased pressure.

REVIEW

1. According to the kinetic-molecular theory

   a. All atoms are the same size.

   b. All atoms have the same kinetic energy.

   c. All atoms are the same mass.

   d. None of the above.

2. In a gas

   a. The atoms and molecules occupy most of the volume of a container.

   b. The atoms and molecules are in constant random motion.

   c. The atoms and molecules spend most of their time colliding with the walls trying to get out.

   d. The atoms and molecules are the same mass.

3. A hydrogen molecule is

   a. Larger than an O_2 molecule.   c. Smaller than a water molecule.
   b. Smaller than a hydrogen atom.   d. Endothermic.

4. In the kinetic-molecular theory, the elastic collision of atoms and molecules can be compared to a

   a. Pool game.   c. Billiard game.
   b. Game of marbles.   d. All of these.

5. The average kinetic energy of an atom or molecule is related to

   a. Size.   c. Atomic number.
   b. Mass.   d. Temperature.

6. The kinetic-molecular theory applies to

   a. Gases.   c. Liquids.
   b. Solids.   d. All of these.

7. When a water molecule evaporates from a container of water, the velocity of the escaping water molecule

   a. Is faster than average.
   b. Is slower than average.
   c. Is the same as the bulk of the water.
   d. Carries more energy than the average.

8. At any given temperature, the pressure exerted by a liquid-gas equilibrium is called

   a. Atmospheric pressure.   c. Gas pressure.
   b. Vapor pressure.   d. None of these.

9. The melting point of a solid is

   a. The temperature at which the solid and gas phase are in equilibrium.
   b. The temperature at which molecular motion stops.
   c. The point where forces of repulsion in the liquid are overcome.
   d. Numerically equal to the freezing point of the liquid.

10. A solid which passes directly from the solid state to the gaseous state is said to

   a. Sublime.   c. Boil.
   b. Evaporate.   d. Vaporize.

11. When a liquid is confined in a closed container

   a. The rate of condensation equals the rate of evaporation.
   b. The atoms or molecules collide with the walls of the container more than if the container were open.
   c. The liquid evaporates slower.
   d. The atoms or molecules collide less frequently with the walls of the container.

12. The rate of a chemical reaction increases when

    a. The concentration of reactants is increased.
    b. The pressure is increased.
    c. The temperature is increased.
    d. All of the above.

13. When water is boiled on a mountain top

    a. The vapor pressure at 212°F is lower than at sea level.
    b. The boiling point of water is 212°F.
    c. The boiling point of water is lower than 212°F.
    d. The boiling point of water is the same as at sea level.

14. When a liquid evaporates, the atoms or molecules which escape

    a. Are the hottest from the bottom.
    b. Are mostly from the edges.
    c. Come randomly from throughout the liquid.
    d. Are from the surface layer.

Unit 6 Gas Laws Governing Temperature, Pressure, and Volume

Anyone who has ever inflated a tire with a hand pump will recall that increasing the pressure on a given volume of gas decreases its volume. When pressure increases, volume decreases. Volume is inversely proportional to pressure.

$$V \, \alpha \, \frac{1}{P}$$ where V is volume

P is pressure

α is the symbol for proportional

This law which relates the effect of a change in pressure on the volume of a gas is called *Boyle's Law.*

Similarly when a closed container is heated, the gas tries to expand. When the temperature increases the volume increases. Volume is directly proportional to temperature.

$$V \, \alpha \, T$$ where T is temperature in $^\circ$K

This law which relates the effect of a change in temperature on the volume of a gas is called *Charles' Law.*

Boyle's Law and Charles' Law may be combined into a general Gas Law.

$$\frac{PV}{T} = \frac{P'V'}{T'}$$ where P, V, T are the initial pressure, volume, and temperature in $^\circ$K.

and P', V', T' are the final pressure, volume, and temperature in $^\circ$K.

This general gas law permits calculation of quantities such as the pressure generated inside of a 55-gallon drum of gasoline when it is exposed to an elevated temperature or the cooling level to be expected when a given volume of gas is expanded. The understanding of such laws is of obvious advantage to personnel involved with the safe handling of hazardous materials.

The calculation method is straightforward and involves only simple arithmetic. The most difficult part relates to the requirement that temperature must be expressed according to the Kelvin temperature scale in $^\circ$K. A conversion table is given in table 6-1.

Temperature in		
$^\circ$F	$^\circ$C	$^\circ$K
−460	−273	0
−418	−250	23
−328	−200	73
−238	−150	123
−148	−100	173
−58	−50	223
32	0	273
122	50	323
212	100	373
302	150	423
392	200	473
572	300	573
752	400	673
932	500	773
1112	600	873
1292	700	973
1472	800	1073
1652	900	1173
1832	1000	1273

Note: Temperatures not listed can be calculated or interpolated by the relationships:

1°C = 1.8°F Each 1°C unit is equal to 1.8°F units or

F = 9/5C + 32

C = 5/9 (F − 32) and

$^\circ$K = $^\circ$C + 273

Table 6-1. Fahrenheit, Centigrade, Kelvin
Temperature Conversion Chart.

EXAMPLE 1

If a gas occupies a volume of 5 cubic feet (ft^3) at a temperature of 122°F and a pressure of 1 atmosphere (atm), what is its volume at a pressure of 1.5 atm and a temperature of 302°F?

$$\frac{PV}{T} = \frac{P'V'}{T'}$$

$P = 1$ atm $\quad P' = 1.5$ atm
$V = 5$ ft³ $\quad V' = ?$
$T = 122°F \quad T' = 302°F$

Step 1: Convert temperatures to °K. (Use table 6-1.)

$T = 122°F = 323°K;$

$T' = 302°F = 423°K$

Step 2: Substitute quantities into the equation.

$$\frac{(1)(5)}{(323)} = \frac{(1.5)(V')}{(423)}$$

Step 3: Cross-multiply and divide.

$(323)(1.5)(V') = (1)(5)(423)$

$(484.5)(V') = (2115)$

$V' = \frac{(2115)}{(484.5)}$

$V' = 4.36$ ft³

EXAMPLE 2

A 55-gallon drum of gasoline is exposed to a high heat source in the winter. If the gasoline vapor occupies a volume of 7.35 cubic feet (the volume of the 55-gallon drum) at a pressure of 1 atmosphere and a temperature of 35°F, what pressure is generated inside the drum when it is heated to 218°F?

$$\frac{PV}{T} = \frac{P'V'}{T'}$$

where $P = 1$ atm $\quad P' = ?$
$\qquad V = 7.35$ ft³ $\quad V' = 7.35$ ft³
$\qquad T = 35°F \quad T' = 218°F$

Step 1: Convert temperatures to °K.

NOTE: Since T = 35°F is not listed in the table, it must be calculated by the relationship 1°C = 1.8°F. Therefore, F = 9/5 C + 32; 35 = 9/5 C + 32; C = 1.7°; 1.7° C + 273 = 274.7°K.
Similarly, since T = 218°F is not listed in the table: F = 9/5 C + 32; 218 = 9/5 C + 32; C = 103.4; 103.4° C + 273 = 376.4°K.

Step 2: Substitute into equation.

$$\frac{(1)(7.35)}{(274.7)} = \frac{(P')(7.35)}{(376.4)}$$

Step 3: Cross-multiply and divide.

$(274.7(7.35)(P') = (1)(7.35)(376.4)$

$(274.7)(P') = (376.4)$

$P' = 1.4$ atmospheres

A knowledge of the gas laws as presented here is often useful in predicting the hazards associated with exposure of pressure vessels to elevated temperatures. It is also part of the fundamental background which is valuable in the interpretation of explosive phenomena which is to be discussed shortly.

REVIEW

1. Boyle's Law states that
   a. Volume is inversely proportional to pressure.
   b. Pressure decreases as volume increases.
   c. Volume decreases as pressure increases.
   d. All of the above.

2. Charles' Law states that

   a. Volume is directly proportional to temperature.
   b. Volume is directly proportional to pressure.
   c. Volume is indirectly proportional to temperature.
   d. Volume is indirectly proportional to pressure.

3. In the General Gas Law temperature is expressed in degrees

   a. Fahrenheit.   c. Kelvin.
   b. Centigrade.   d. Any of these.

4. 141°C is numerically equal to

   a. 341.0°F and 441°K.   c. 386.5°F and 562°K.
   b. 285.8°F and 414°K.   d. None of these.

5. 1103°K is numerically equal to

   a. 595°C and 868°K.   c. 868°C and 695°K.
   b. 695°C and 978°K.   d. None of these.

6. 1000°K is numerically equal to

   a. 841°C and 1527°F.   c. 727°C and 1341°F.
   b. 1341°C and 727°F.   d. None of these.

7. If a gas occupies a volume of 2.5 ft³ at a temperature of 1220°F and a pressure of 1 atmosphere, what is the volume at a pressure of 1.5 atm and 302°F?

   a. 0.1 ft³   c. 0.8 ft³
   b. 0.3 ft.   d. 0.2 ft³

8. If a gas occupies a volume of 10 ft³ at a pressure of 500 atm and a temperature of 68°F, what is the pressure generated at a temperature of 734°F?

   a. 1131 atm.   c. 1567 atm.
   b. 756 atm.   d. 1328 atm.

Unit 7 Explosion Mechanisms

Much of the confusion and uncertainty about explosives stems from the fact that many so-called definitions are really descriptive terms that classify various types of rapid energy-releasing processes. There is really only one definition! All explosions, viewed in terms of the kinetic-molecular theory, proceed according to identical mechanisms. The student who understands how one explosive works can understand how all explosives work!

COMBUSTION VS EXPLOSION

First of all, it is necessary to realize the difference between a combustion and an explosion. The fundamental difference is shown in figure 7-1. As a combustion proceeds, the rate of energy release is balanced by energy dissipation, such that a limiting rate of reaction is reached. The much-used example of the combustion of methane in air to produce carbon dioxide and water shows that when methane is burned,

$$CH_4 + 2O_2 \rightarrow CO_2 + 2H_2O + \text{heat}$$

the combustion products expand away from the combustion zone taking with them kinetic energy in the form of heat. The plateau level of the combustion curve is established because a limit is placed on the temperature and pressure of the system. The rate of reaction reaches an equilibrium value.

However, if the methane and oxygen are placed in a closed container and the reaction is initiated, an explosion results. In this case, the only difference is that the products of combustion and, therefore, the heat are not removed from the reaction zone. The temperature continues to rise. Since the system is confined, the pressure rises also. Both effects operate in the direction of increasing the rate of reaction. There is no mechanism for dissipation of energy; therefore, the rate of reaction increases exponentially until all of the reactants are consumed.

Some examples of common explosive reactions are

- Rapid oxidations such as the combustion of nitroglycerin, nitrocellulose, and ammonium nitrate.

- Rapid oxidations such as gasoline-air in a closed container.

- Rapid thermal decomposition of materials such as hydrogen peroxide and hydrazine.

- Miscellaneous situations such as chemical reactions which are out of control.

- Miscellaneous situations such as pressure vessel failure.

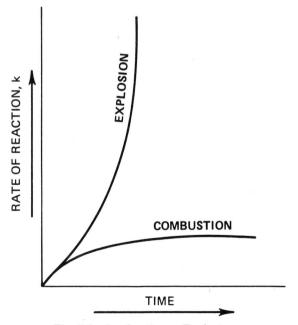

Fig. 7-1. Combustion vs Explosion.

Nearly everyone is familiar with explosive materials such as nitroglycerin and nitrocellulose. Added insight into the nature of explosions can be gained by considering why nitroglycerin and nitrocellulose are explosives. Chemical composition and bond energies are the reasons! Nitroglycerin, for example, is composed of carbon, oxygen, nitrogen, and hydrogen. The empirical formula for nitroglycerin is $C_3H_5N_3O_9$ and the structural formula is

```
          H
          |
     H–C–O–NO2
          |
     H–C–O–NO2
          |
     H–C–O–NO2
          |
          H
```

The empirical formula indicates the number of C, H, N, O atoms: 3 carbon atoms, 5 hydrogen atoms, 3 nitrogen atoms, and 9 oxygen atoms. However, the empirical formula does not tell how the atoms are bonded together. The exact bonding of atoms is shown by the structural formula.

The explosion of nitroglycerin results in the same reaction products as the combustion of the compound. An explosion simply produces them at a much faster rate. The equation for the breakdown of nitroglycerin to carbon dioxide, water, and nitrogen shows that it is unbalanced by only one oxygen

$$2C_3H_5N_3O_9 \rightarrow 6CO_2 + 5H_2O + 3N_2$$

atom. An explosive like nitroglycerin contains sufficient oxygen together with the other atoms to oxidize or burn without recourse to atmospheric oxygen. This is the reason why explosives can function in closed containers, underwater, or in outer space. They carry their own oxygen.

The other requirement of an explosive is that when the molecule is rearranged into its reaction products, heat must be liberated. It must be an exothermic reaction. All homogeneous explosives possess these two characteristics in varying degrees:

- They carry much of their own oxidizer.
- They are exothermic.

The relative hazard or explosiveness of a wide variety of materials can be estimated by these characteristics. They vary primarily in their degree of exothermicity. Rapid oxidations such as those encountered with gasoline-oxygen mixtures are explosive if enough oxygen is premixed with the gasoline vapor to permit an essentially complete reaction. The reaction becomes explosive because the oxygen atoms are close to the carbon and hydrogen atoms and can therefore react instantaneously. In other circumstances, gasoline simply burns because the rate of the reaction is controlled by the diffusion of oxygen to the combustion zone.

Explosion of materials such as hydrogen peroxide and hydrazine is a type of rapid oxidation reaction. Both peroxide and hydrazine have the property that they decompose exothermically (with the evolution of heat).

(Hydrogen Peroxide) $2H_2O_2 \rightarrow 2H_2O + O_2 + heat$

(Hydrazine) $2N_2H_4 \rightarrow N_2 + H_2 + 2NH_3 + heat$

Often hazardous situations are encountered which were not intended to be that way. Explosives are inherently hazardous because of their intended function. Many chemical reactions can be potentially explosive if their controlling mechanism fails. If the heat buildup proceeds too rapidly, the reaction rate escalates, and an explosion can result. Many examples of this type are discussed throughout the text for specific situations.

A pressure vessel failure, such as the explosion of a water heater, may be considered a physical explosion rather than a chemical explosion. The end-product of all explosions is a pressure wave. In a water heater explosion, the pressure wave is caused by vaporizing water. When the strength of the pressure

$$H_2O \text{ (liquid)} \rightarrow H_2O \text{ (gas)}$$

vessel is exceeded, an unconfined pressure wave is generated.

It should be pointed out, however, that every molecule which contains oxygen is not an explosive. Sodium carbonate (Na_2CO_3) is not explosive even though it contains a high percentage of oxygen. In fact, it is endothermic and withdraws heat in order to dissociate.

Many chemical compounds (the sulfates, phosphates, borates, for example) fall into this category because they are endothermic. Some of them do have toxicity hazards even though they do not have explosive hazards.

REVIEW

1. In the unconfined combustion of methane and oxygen as represented in figure 7-1, the plateau level is established because
   a. The rate of reaction reaches an equilibrium value.
   b. The gases are heated.
   c. The pressure of the gases is higher.
   d. All of the above.

2. In the unconfined combustion of methane and oxygen as represented in figure 7-1,
   a. The temperature reaches a limiting value.
   b. The pressure reaches a limiting value.
   c. The combustion products carry away kinetic energy.
   d. All of the above.

3. In the confined combustion of methane and oxygen as represented in figure 7-1,
   a. The rate of reaction is increased.
   b. An explosion results.
   c. The kinetic energy level is increased.
   d. All of the above.

4. In the confined combustion of methane and oxygen as represented in figure 7-1,
   a. Heat is not removed from the reaction zone.
   b. The temperature rises.
   c. The pressure rises.
   d. All of the above.

5. A fundamental difference between an explosion and a combustion is that in an explosion
   a. The effects of temperature and pressure operate in the direction of increasing the rate of reaction.
   b. Energy is dissipated more efficiently.
   c. Heat is removed from the reaction zone more quickly.
   d. All of the above.

6. Nitroglycerin is an explosive because

   a. It contains the elements C, H, O, N.
   b. It carries its own oxygen.
   c. It is exothermic.
   d. It is endothermic.

7. Sodium carbonate (Na_2CO_3)

   a. Is an explosive.   c. Is exothermic.
   b. Carries its own oxygen.   d. None of these.

8. Although all explosives are exothermic materials, endothermic materials can also be hazardous because

   a. They burn readily.   c. They are reactive with water.
   b. They have high flash points.   d. They can be toxic.

9. A water heater explosion and a nitrocellulose explosion are similar in that they both

   a. Generate a pressure wave.   c. Are exothermic.
   b. Generate heat.   d. Have rapid chemical reactions.

Unit 8 Shock Waves and Explosions

Another aspect of explosive behavior related to detonation also warrants discussion. It is discussed under this separate heading because it is a physical rather than a chemical phenomenon.

Some explosives like nitrocellulose which are highly energetic tend to burn intensely (*deflagrate*) but they do not usually detonate. Others such as nitroglycerin have a great tendency to detonate. Explosives of this type are classified as *brisant explosives.* The shattering effect of brisance, an effect due to detonation of an explosive, is caused by the detonation wave (shock wave) of the explosive. A shock wave generated by a supersonic aircraft is called a sonic boom. Its origin and hazardous effects are similar to the shock wave generated by an explosive.

When an aircraft is in flight, the pressure waves caused by the compression of air formed at the nose and wing edges expand to lower density regions. At speeds less than the velocity of sound, the compression waves have time to expand away from the aircraft as shown in figure 8-1. In contrast, when the aircraft is flying faster than the compression waves, the waves are compressed to extremely high pressure levels. Associated with the rapid compression is an intense heating effect. The high pressure-high temperature wave expands violently creating the shock wave commonly referred to as a sonic boom. Even though most of the energy of a sonic boom is dissipated before it reaches the ground, but even at that, it often causes extensive damage.

A similar situation can arise with explosives. When an explosive is ignited, the first step in the complex sequence is a combustion process with the resultant generation of gases. If the combustion is rapid enough, the combustion gases are produced at a rate faster than they can expand away from the reaction

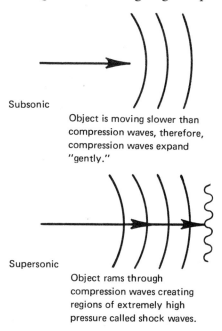

Fig. 8-1. Shock Waves and Sonic Booms.

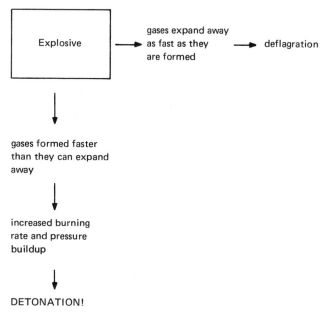

Fig. 8-2. Conditions Which Determine Whether An Explosive Burns Or Detonates.

site. When this happens, a compression wave is formed within the explosive, which is analogous to the compression wave formed by a supersonic aircraft.

The second step in this type of explosion process is detonation. When the compression wave or shock wave is formed, the remainder of the explosive is in effect "ignited" at an extremely high pressure and temperature. Typical shock wave temperatures are 5000°C (9032°F) and typical pressures are thousands of pounds per square inch. The remaining portion of the explosive, therefore, detonates.

The sequence of events which determine whether an explosive burns normally or detonates are summarized in figure 8-2.

REVIEW

1. A shock wave is an example of a
    a. Chemical process.
    b. Physical process.
    c. Combustion process.
    d. Exothermic process.

2. A shock wave is generated in
    a. A detonation.
    b. An explosion.
    c. A rapid combustion.
    d. A subsonic aircraft.

3. The first step which occurs when an explosive such as nitroglycerin is ignited is that
    a. A shock wave is generated.
    b. It explodes.
    c. Gases are generated.
    d. All of these.

4. All explosives
    a. Detonate.
    b. Deflagrate.
    c. Generate shock waves.
    d. Are endothermic.

5. A shock wave can be compared to
    a. An ocean wave.
    b. A sound wave.
    c. A light wave.
    d. A compression wave.

6. A detonation occurs when
    a. Gases are formed faster than they can expand away.
    b. Gases expand away as fast as they are formed.
    c. Large quantities of gas are generated.
    d. Exothermic materials are ignited quickly.

7. A detonation is more hazardous than an explosion because of the
    a. Higher temperature.
    b. Higher pressure.
    c. Higher burning rate.
    d. Shock waves.

8. A shock wave generated by a supersonic aircraft is different than that generated by detonation of an explosive because
    a. It is physical in origin rather than chemical.
    b. It occurs at supersonic velocities.
    c. It occurs at higher altitudes.
    d. None of the above.

Unit 9 Toxicity

Toxicity deals with the effects of a poisonous material when absorbed into the body via swallowing, breathing, or skin absorption. It implies a knowledge of the source, the symptoms, the physiological effect, and remedial measures. The primary concern with toxicity in this text is from a preventative point of view: a concern with the types of toxic substances which are apt to be encountered in various situations related to fire protection, explosives, and hazardous materials.

The purpose of this unit is to survey the potential occurrence of toxic hazardous materials in a very general way and to define some of the vocabulary used throughout the text. Later, in section 7, the physiological effects, symptoms, and preventative measures applicable to a wide range of hazardous materials are discussed.

All fire protection personnel are vitally concerned with the indirect formation of toxic combustion products from normally nontoxic materials. For example, everyone is familiar with the formation of the toxic gas, carbon monoxide, from the incomplete combustion of hydrocarbons. In the terminology used in this text, this often is referred to as a CHO (carbon-hydrogen-oxygen) system. The combustion of a CHO system can result only in the formation of carbon dioxide, water, carbon monoxide, and carbon soot because these are the only products possible from a system containing only carbon, hydrogen, and oxygen.

In the combustion of polyvinylchloride plastic or a chlorine gas spill, two additional toxic gases are formed: hydrogen chloride (HCl) and small amounts of phosgene ($COCl_2$). A fundamental background is needed to understand the implications. The formation of hydrogen chloride (HCl) and phosgene ($COCl_2$) in the combustion of a carbon, hydrogen, oxygen, chlorine (CHOCl) system is shown in table 9-1 together with toxicity ratings in parts per million (ppm). It can be seen, for example, that HCl is 10 times as toxic as CO. Phosgene ($COCl_2$) is 500 times as toxic. Phosgene has the odor of freshly mowed grass; it is a gas that, when inhaled, produces a heavy feeling in the throat. The quantities of HCl and $COCl_2$, although small, can be significant in a fire situation involving chlorinated plastics, chlorinated solvents, or chlorine.

Systems containing fluorine (F) or bromine (Br) form the corresponding fluoride and bromide compounds. Fluorine toxic products are most apt to be encountered with fluorine spills or combustion of fluorine-containing plastics. Bromine toxic products are even less likely to be encountered except in the case of bromine spills, but are included for the sake of completeness.

Although the data for chlorine, fluorine, and bromine may seem a bit detailed and perhaps too great to retain, a very simple generalization can be made. For uncontrolled

System	Nontoxic Products	Toxic Products	Threshold Toxicity Limits ppm
CHO	CO_2, H_2O	CO	50
CHOCl	CO_2, H_2O	HCl	5
		$COCl_2$	0.1
CHON	CO_2, H_2O, N_2	HCN	10
CHOS	CO_2, H_2O	SO_2	5
CHS	CO_2	H_2S	10
CHOF	CO_2, H_2O	HF	3
		COF_2	0.1
CHOBr	CO_2, H_2O	HBr	3
		$COBr_2$	0.1

Table 9-1. Toxic Combustion Products Likely To Be Encountered in Fire Protection Situations.

fire situations in which the concentration is largely unknown, all halogen (F, Cl, Br, I) fires can be considered to yield toxic products, which are much more toxic than carbon monoxide.

Similarly, fires containing nitrogen compounds are potentially very hazardous because all C, H, O, N systems form hydrogen cyanide (HCN) as a toxic combustion product. For reasons that are not entirely apparent, everyone seems to know that HCN is very toxic, perhaps because of certain applications formerly in vogue among some law agencies. However, inspection of table 9-1 shows that HCN is not as toxic as HCl. It is potentially more hazardous, however, because it is odorless whereas HCl has a pungent, acid odor.

One very common source of CHON fires is in the combustion of carpets. Both wool and nylon are high in nitrogen and therefore produce high concentrations of toxic HCN combustion products. Other potential sources of HCN are fires involving polyurethane foam, food products and meat products.

Sulfur-containing materials form sulfur dioxide (SO_2) in systems containing oxygen and hydrogen sulfide in systems lacking in oxygen. Both are highly hazardous toxic combustion products.

This brief background provides a basis for discussion of specific hazardous materials in succeeding sections. In any discussion of toxicity, it is also necessary to have some understanding of the various methods of rating the toxicity of a given substance.

- Exposure may be acute or chronic.

    1. *Acute* means a single dose or exposure.

    2. *Chronic* means repeated exposure.

    3. *Subacute* means an exposure two or three times at a level between acute and chronic.

- Exposure may be local or systemic.

    1. *Local* means action at the site of contact: A caustic material such as lye attacks at the site of contact.

    2. *Systemic* means the toxic material must pass through the skin, mucous membranes, or lung and be transported to the site of action via the bloodstream: Chlorinated solvents are a familiar example of a systemic poison. They are primarily liver poisons. They must be absorbed and transported to the liver in the bloodstream.

- Toxic materials are often rated in the following terms.

    1. Slight, moderate, severe.

    2. Acute local, acute systemic, chronic local, chronic systemic.

    3. LD_{50} means dosage (ordinarily expressed in mgm/kg. body weight) at which 50% fatality results.

    4. MLD means minimum lethal dosage (mgm/kg. body weight) at which any fatalities result.

    5. TL means threshold limit. This was formerly called the maximum allowable concentration (MAC) and is the dangerous level for exposure for a standard eight-hour work day.

Specific toxicity ratings (TL, MLD, LD_{50}) are tabulated with the more detailed discussion of various classes of materials. Each rating has some advantages and many disadvantages. Fire fighters, for example, may be less concerned with threshold limits for long-term exposure than with the minimum lethal dosage (MLD).

Toxicity ratings are usually not available in all of the systems. All too commonly, toxicity ratings are available in only one system or not at all. For this reason, it is considered important to have at least a minimal knowledge of the various toxicity definitions and rating systems.

REVIEW

1. The most commonly encountered toxic combustion product in systems containing only the elements carbon, hydrogen, and oxygen is
   a. Ozone, O_3.
   b. Carbon dioxide.
   c. Carbon monoxide.
   d. Carbon.

2. Toxic combustion products commonly encountered in the combustion of chlorinated plastics are
   a. Chlorine and carbon monoxide.
   b. Chlorine and phosgene.
   c. Phosgene and carbon monoxide.
   d. Phosgene and hydrogen chloride.

3. The empirical formula for phosgene is
   a. COCl.
   b. CO_2Cl.
   c. $COCl_2$.
   d. C_2OCl.

4. Which of the following combustion products is the most toxic?
   a. Carbon monoxide.
   b. Phosgene.
   c. Hydrogen cyanide.
   d. Hydrogen sulfide.

5. The toxic combustion product, HF, is most likely to be encountered in the combustion of
   a. Polyethylene.
   b. Polyurethane.
   c. Carpets.
   d. Teflon®.

6. The toxic combustion product, HCN, may be encountered in the combustion of
   a. Polyurethane.
   b. Carpets.
   c. Meat products.
   d. Any of these.

7. In a fire situation in which CO is formed because of limited oxygen, which of the additional toxic products would be likely to form under the same conditions?
   a. H_2S.
   b. SO_2.
   c. NO_2.
   d. All of these.

8. In the comparison of the relative toxicity of two substances
   a. Acute toxicity is worse than chronic toxicity.
   b. Chronic toxicity is worse than acute toxicity.
   c. Chronic means repeated exposure.
   d. None of the above.

9. The abbreviation LD as in LD_{50} means
   a. Lethal dosage.
   b. Likely dosage.
   c. Light dose.
   d. None of these.

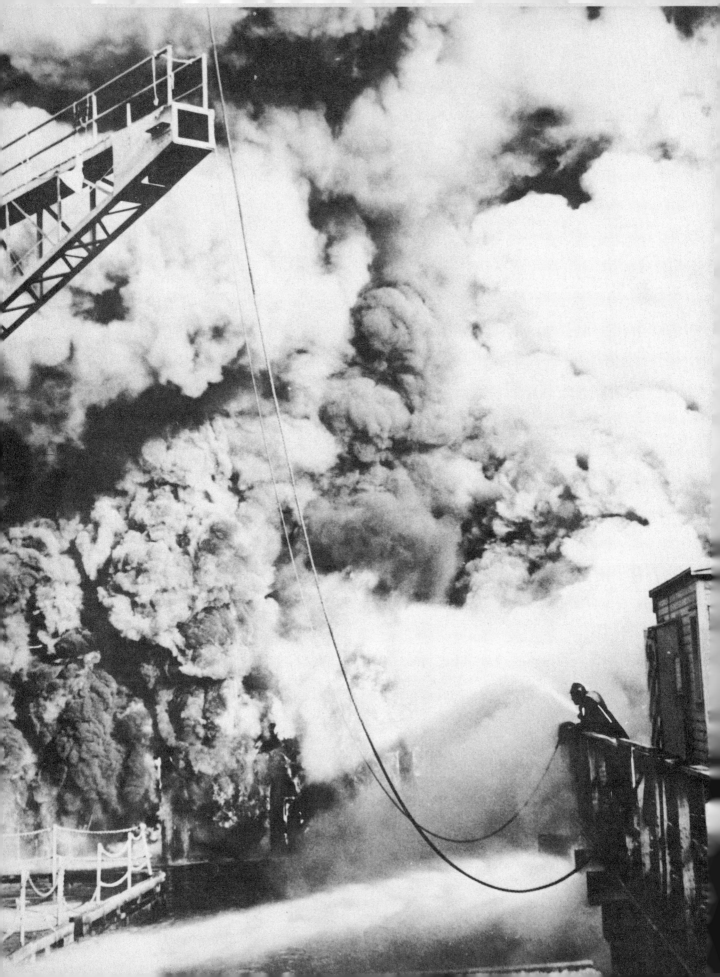

SECTION 2

COMBUSTION

Unit 10 Physical Properties Which Affect Combustion Behavior

Consideration of a relatively small number of physical properties is sufficient to aid understanding and prediction of the combustion behavior of most materials. Examples of physical properties are color, density, melting point, and boiling point[1], as compared to chemical properties, such as reactivity with oxygen and water, which are considered later.

FLAMMABILITY LIMITS

The combustion of kerosene in oxygen is a typical situation. Kerosene has a melting point of approximately $-22°F$ and a boiling point of approximately $662°F$. When kerosene is in the solid state at temperatures below $-22°F$, the vapor pressure of the gas above the solid is very low. It does not ignite when an ignition source is applied because the amount of kerosene vaporized is low. When the temperature is raised to approximately $110°F$, application of an ignition source results in combustion. The minimum temperature at which the vapor pressure above a liquid or solid is high enough to result in ignition is defined as the *flash point*. Each material also has a characteristic *lower flammable limit* which is the minimum volume percent of the material in air which can be ignited. For kerosene, this value is 0.7 percent.

Each material also has an *upper flammable limit* which is the maximum volume percent of the material in air which can be ignited. For kerosene, this value is 5 percent. The significance of the upper flammable limit is not always as obvious as that of the lower flammable limit. If the upper flammable limit is exceeded, the mixture cannot be ignited and sustain combustion. The balanced chemical equation for the combustion of kerosene is shown:

$$C_{11}H_{24} + 17\,O_2 \rightarrow 11\,CO_2 + 12\,H_2O$$

If equal volumes of gases contain equal numbers of molecules (which is true!), the percentage of kerosene in air under *stoichiometric combustion conditions* (the system is balanced as in the balanced chemical equation) is 1/17 of the total volume or 5.5 volume percent. The upper flammability limit is approximately equal to the relative numbers of kerosene and oxygen atoms indicated in the balanced chemical equation. If the proportion of kerosene molecules is increased, insufficient oxygen is present. Similarly, if the proportion of kerosene molecules is decreased, insufficient kerosene is present to sustain combustion. Stable combustion of kerosene is sustained only within a narrow range of concentration limits.

The sequence of processes which occur with any material are the same as those described for kerosene-air. Gasoline-air is basically the same except for the fact that the flash point is approximately $-40°F$. The lower flammable limit is 1.4 volume percent and the upper flammable limit is 7.5%. A comparison of the similarities and differences of the interrelationship between physical properties and combustion behavior for kerosene and gasoline is summarized in figure 10-1, page 36.

---

[1]Since the fire service will eventually be affected by the change to metric measurement, a conversion table is included for the convenience of the student.

FLASH POINT

The primary physical property parameters involved in determining the combustion behavior of gasoline and kerosene are flammability limits and flash point. Other factors such as vapor density, specific gravity, water solubility, ignition temperature, and thermal decomposition temperature are often important. However, they are generally of secondary importance.

The concept of flash point is equally valid for liquids, gases, and solids. For a gas, the volume percent in air of the material is not limited or indirectly determined by vaporization. As shown in figure 10-2, whenever the concentration of the gas in air exceeds the lower flammable limit, ignition can occur. A gas is always above the flash point, as it has been defined. This is one of the primary reasons for the increased fire hazard associated with a flammable gas. Whenever the concentration limit exceeds the lower flammable limit, ignition can occur.

For a solid, such as wax, or a thermoplastic, such as cellulose acetate, the sequence of events is similar to that of a liquid.

- The solid is heated toward the melting point. If the vapor pressure of the solid is sufficiently high to yield a concentration of vapors exceeding the lower flammable limit, ignition can occur. The common definition of flash point is applicable. For most materials of this type, however, melting occurs before the lower flammability level is reached.

- The solid melts to yield a liquid. The definition of flash point applies as for any liquid.

Note: *Thermoplastics* are plastics with little or no cross-linking between polymer-chains. Because they are not cross-linked, they flow or melt when exposed to heat.

Other types of solids such as wood (which is a highly cross-linked polymer and remains tightly bonded to the point of thermal breakdown) break down to yield gaseous products when heated. Wood does not pass through a liquid phase during thermal decomposition. Whenever the vapor concentration

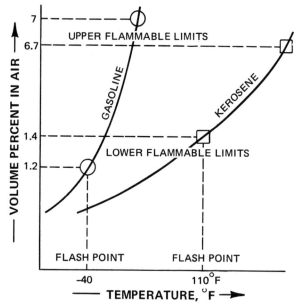

Fig. 10-1. Physical Steps Involved in the Combustion Behavior of Kerosene-Air and Gasoline-Air.

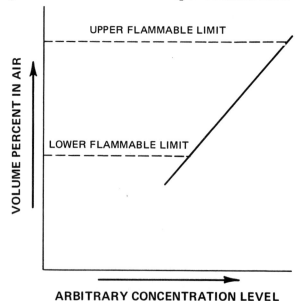

Fig. 10-2. Application of Flammable Limit Concepts to a Gaseous System.

The values in the body of the table give, in degrees Fahrenheit, the temperatures indicated in degrees Centigrade at the top and side.

$1°C. = 1.8°F.$
For temperatures below $0°C.$

Temp. °C.	0	1	2	3	4	5	6	7	8	9
0	+32.0	30.2	28.4	26.6	24.8	23.0	21.2	19.4	17.6	15.8
-10	+14.0	12.2	10.4	8.6	6.8	5.0	3.2	+1.4	-0.4	-2.2
-20	-4.0	5.8	7.6	9.4	11.2	13.0	14.8	16.6	18.4	20.2
-30	-22.0	23.8	25.6	27.4	29.2	31.0	32.8	34.6	36.4	38.2
-40	-40.0	41.8	43.6	45.4	47.2	49.0	50.8	52.6	54.4	56.2
-50	-58.0	59.8	61.6	63.4	65.2	67.0	68.8	70.6	72.4	74.2
-60	-76.0	77.8	79.6	81.4	83.2	85.0	86.8	88.6	90.4	92.2
-70	-94.0	95.8	97.6	99.4	101.2	103.0	104.8	106.6	108.4	110.2
-80	-112.0	113.8	115.6	117.4	119.2	121.0	122.8	124.6	126.4	128.2
-90	-130.0	131.8	133.6	135.4	137.2	139.0	140.8	142.6	144.4	146.2
-100	-148.0	149.8	151.6	153.4	155.2	157.0	158.8	160.6	162.4	164.2
-110	-166.0	167.8	169.6	171.4	173.2	175.0	176.8	178.6	180.4	182.2
-120	-184.0	185.8	187.6	189.4	191.2	193.0	194.8	196.6	198.4	200.2
-130	-202.0	203.8	205.6	207.4	209.2	211.0	212.8	214.6	216.4	218.2
-140	-220.0	221.8	223.6	225.4	227.2	229.0	230.8	232.6	234.4	236.2
-150	-238.0	239.8	241.6	243.4	245.2	247.0	248.8	250.6	252.4	254.2
-160	-256.0	257.8	259.6	261.4	263.2	265.0	266.8	268.6	270.4	272.2
-170	-274.0	275.8	277.6	279.4	281.2	283.0	284.8	286.6	288.4	290.2
-180	-292.0	293.8	295.6	297.4	299.2	301.0	302.8	304.6	306.4	308.2
-190	-310.0	311.8	313.6	315.4	317.2	319.0	320.8	322.6	324.4	326.2
-200	-328.0	329.8	331.6	333.4	335.2	337.0	338.8	340.6	342.4	344.2
-210	-346.0	347.8	349.6	351.4	353.2	355.0	356.8	358.6	360.4	362.2
-220	-364.0	365.8	367.6	369.4	371.2	373.0	374.8	376.6	378.4	380.2
-230	-382.0	383.8	385.6	387.4	389.2	391.0	392.8	394.6	396.4	398.2
-240	-400.0	401.8	403.6	405.4	407.2	409.0	410.8	412.6	414.4	416.2
-250	-418.0	419.8	421.6	423.4	425.2	427.0	428.8	430.6	432.4	434.2
-260	-436.0	437.8	439.6	441.4	443.2	445.0	446.8	448.6	450.4	452.2
-270	-454.0	455.8	457.6	459.4	. . . . .	. . . . .	. . . . .	. . . . .	. . . . .	. . . . .

For temperatures above $0°C.$

Temp. °C.	0	1	2	3	4	5	6	7	8	9
0	32.0	33.8	35.6	37.4	39.2	41.0	42.8	44.6	46.4	48.2
10	50.0	51.8	53.6	55.4	57.2	59.0	60.8	62.6	64.4	66.2
20	68.0	69.8	71.6	73.4	75.2	77.0	78.8	80.6	82.4	84.2
30	86.0	87.8	89.6	91.4	93.2	95.0	96.8	98.6	100.4	102.2
40	104.0	105.8	107.6	109.4	111.2	113.0	114.8	116.6	118.4	120.2
50	122.0	123.8	125.6	127.4	129.2	131.0	132.8	134.6	136.4	138.2
60	140.0	141.8	143.6	145.4	147.2	149.0	150.8	152.6	154.4	156.2
70	158.0	159.8	161.6	163.4	165.2	167.0	168.8	170.6	172.4	174.2
80	176.0	177.8	179.6	181.4	183.2	185.0	186.8	188.6	190.4	192.2
90	194.0	195.8	197.6	199.4	201.2	203.0	204.8	206.6	208.4	210.2
100	212.0	213.8	215.6	217.4	219.2	221.0	222.8	224.6	226.4	228.2
110	230.0	231.8	233.6	235.4	237.2	239.0	240.8	242.6	244.4	246.2
120	248.0	249.8	251.6	253.4	255.2	257.0	258.8	260.6	262.4	264.2
130	266.0	267.8	269.6	271.4	273.2	275.0	276.8	278.6	280.4	282.2
140	284.0	285.8	287.6	289.4	291.2	293.0	294.8	296.6	298.4	300.2
150	302.0	303.8	305.6	307.4	309.2	311.0	312.8	314.6	316.4	318.2
160	320.0	321.8	323.6	325.4	327.2	329.0	330.8	332.6	334.4	336.2
170	338.0	339.8	341.6	343.4	345.2	347.0	348.8	350.6	352.4	354.2
180	356.0	357.8	359.6	361.4	363.2	365.0	366.8	368.6	370.4	372.2
190	374.0	375.8	377.6	379.4	381.2	383.0	384.8	386.6	388.4	390.2

$-273.16°C = -459.72°F.$ = absolute zero

For	°C	0.1	0.2	0.3	0.4	0.5	0.6	0.7	0.8	0.9	1.0
interpolation	°F	0.18	0.36	0.54	0.72	0.90	1.08	1.26	1.44	1.62	1.80

Table 10-1. English-Metric Temperature Conversion (continued on page 38).

For temperatures above 0°C.

Temp. °C.	0	1	2	3	4	5	6	7	8	9
200	392.0	393.8	395.6	397.4	399.2	401.0	402.8	404.6	406.4	408.2
210	410.0	411.8	413.6	415.4	417.2	419.0	420.8	422.6	424.4	426.2
220	428.0	429.8	431.6	433.4	435.2	437.0	438.8	440.6	442.4	444.2
230	446.0	447.8	449.6	451.4	453.2	455.0	456.8	458.6	460.4	462.2
240	464.0	465.8	467.6	469.4	471.2	473.0	474.8	476.6	478.4	480.2
250	482.0	483.8	485.6	487.4	489.2	491.0	492.8	494.6	496.4	498.2
260	500.0	501.8	503.6	505.4	507.2	509.0	510.8	512.6	514.4	516.2
270	518.0	519.8	521.6	523.4	525.2	527.0	528.8	530.6	532.4	534.2
280	536.0	537.8	539.6	541.4	543.2	545.0	546.8	548.6	550.4	552.2
290	554.0	555.8	557.6	559.4	561.2	563.0	564.8	566.6	568.4	570.2
300	572.0	573.8	575.6	577.4	579.2	581.0	582.8	584.6	586.4	588.2
310	590.0	591.8	593.6	595.4	597.2	599.0	600.8	602.6	604.4	606.2
320	608.0	609.8	611.6	613.4	615.2	617.0	618.8	620.6	622.4	624.2
330	626.0	627.8	629.6	631.4	633.2	635.0	636.8	638.6	640.4	642.2
340	644.0	645.8	647.6	649.4	651.2	653.0	654.8	656.6	658.4	660.2
350	662.0	663.8	665.6	667.4	669.2	671.0	672.8	674.6	676.4	678.2
360	680.0	681.8	683.6	685.4	687.2	689.0	690.8	692.6	694.4	696.2
370	698.0	699.8	701.6	703.4	705.2	707.0	708.8	710.6	712.4	714.2
380	716.0	717.8	719.6	721.4	723.2	725.0	726.8	728.6	730.4	732.2
390	734.0	735.8	737.6	739.4	741.2	743.0	744.8	746.6	748.4	750.2
400	752.0	753.8	755.6	757.4	759.2	761.0	762.8	764.6	766.4	768.2
410	770.0	771.8	773.6	775.4	777.2	779.0	780.8	782.6	784.4	786.2
420	788.0	789.8	791.6	793.4	795.2	797.0	798.8	800.6	802.4	804.2
430	806.0	807.8	809.6	811.4	813.2	815.0	816.8	818.6	820.4	822.2
440	824.0	825.8	827.6	829.4	831.2	833.0	834.8	836.6	838.4	840.2
450	842.0	843.8	845.6	847.4	849.2	851.0	852.8	854.6	856.4	858.2
460	860.0	861.8	863.6	865.4	867.2	869.0	870.8	872.6	874.4	876.2
470	878.0	879.8	881.6	883.4	885.2	887.0	888.8	890.6	892.4	894.2
480	896.0	897.8	899.6	901.4	903.2	905.0	906.8	908.6	910.4	912.2
490	914.0	915.8	917.6	919.4	921.2	923.0	924.8	926.6	928.4	930.2
500	932.0	933.8	935.6	937.4	939.2	941.0	942.8	944.6	946.4	948.2
510	950.0	951.8	953.6	955.4	957.2	959.0	960.8	962.6	964.4	966.2
520	968.0	969.8	971.6	973.4	975.2	977.0	978.8	980.6	982.4	984.2
530	986.0	987.8	989.6	991.4	993.2	995.0	996.8	998.6	1000.4	1002.2
540	1004.0	1005.8	1007.6	1009.4	1011.2	1013.0	1014.8	1016.6	1018.4	1020.2
550	1022.0	1023.8	1025.6	1027.4	1029.2	1031.0	1032.8	1034.6	1036.4	1038.2
560	1040.0	1041.8	1043.6	1045.4	1047.2	1049.0	1050.8	1052.6	1054.4	1056.2
570	1058.0	1059.8	1061.6	1063.4	1065.2	1067.0	1068.8	1070.6	1072.4	1074.2
580	1076.0	1077.8	1079.6	1081.4	1083.2	1085.0	1086.8	1088.6	1090.4	1092.2
590	1094.0	1095.8	1097.6	1099.4	1101.2	1103.0	1104.8	1106.6	1108.4	1110.2
600	1112.0	1113.8	1115.6	1117.4	1119.2	1121.0	1122.8	1124.6	1126.4	1128.2
610	1130.0	1131.8	1133.6	1135.4	1137.2	1139.0	1140.8	1142.6	1144.4	1146.2
620	1148.0	1149.8	1151.6	1153.4	1155.2	1157.0	1158.8	1160.6	1162.4	1164.2
630	1166.0	1167.8	1169.6	1171.4	1173.2	1175.0	1176.8	1178.6	1180.4	1182.2
640	1184.0	1185.8	1187.6	1189.4	1191.2	1193.0	1194.8	1196.6	1198.4	1200.2
650	1202.0	1203.8	1205.6	1207.4	1209.2	1211.0	1212.8	1214.6	1216.4	1218.2
660	1220.0	1221.8	1223.6	1225.4	1227.2	1229.0	1230.8	1232.6	1234.4	1236.2
670	1238.0	1239.8	1241.6	1243.4	1245.2	1247.0	1248.8	1250.6	1252.4	1254.2
680	1256.0	1257.8	1259.6	1261.4	1263.2	1265.0	1266.8	1268.6	1270.4	1272.2
690	1274.0	1275.8	1277.6	1279.4	1281.2	1283.0	1284.8	1286.6	1288.4	1290.2
700	1292.0	1293.8	1295.6	1297.4	1299.2	1301.0	1302.8	1304.6	1306.4	1308.2
710	1310.0	1311.8	1313.6	1315.4	1317.2	1319.0	1320.8	1322.6	1324.4	1326.2
720	1328.0	1329.8	1331.6	1333.4	1335.2	1337.0	1338.8	1340.6	1342.4	1344.2
730	1346.0	1347.8	1349.6	1351.4	1353.2	1355.0	1356.8	1358.6	1360.4	1362.2
740	1364.0	1365.8	1367.6	1369.4	1371.2	1373.0	1374.8	1376.6	1378.4	1380.2
750	1382.0	1383.8	1385.6	1387.4	1389.2	1391.0	1392.8	1394.6	1396.4	1398.2
760	1400.0	1401.8	1403.6	1405.4	1407.2	1409.0	1410.8	1412.6	1414.4	1416.2
770	1418.0	1419.8	1421.6	1423.4	1425.2	1427.0	1428.8	1430.6	1432.4	1434.2
780	1436.0	1437.8	1439.6	1441.4	1443.2	1445.0	1446.8	1448.6	1450.4	1452.2
790	1454.0	1455.8	1457.6	1459.4	1461.2	1463.0	1464.8	1466.6	1468.4	1470.2

For interpolation	°C	0.1	0.2	0.3	0.4	0.5	0.6	0.7	0.8	0.9	1.0
	°F	0.18	0.36	0.54	0.72	0.90	1.08	1.26	1.44	1.62	1.80

exceeds the lower flammable limit, ignition can occur. The physical process is identical. In order to retain a self-consistent definition for flash point, the term *ignition temperature* is used for solids.

Classification systems based on flash point, sometimes combined with other parameters, are widely used for rating the hazards associated with flammable materials. Depending on the flash point, a material is classified as flammable or combustible. A *flammable liquid* is any liquid having a flash point below 100°F and a vapor pressure not exceeding 40 pounds per square inch absolute at 100°F (37.8°C). Any material with a flash point of 100°F or above is classified as a *combustible*. Gasoline is thereby considered flammable whereas kerosene is defined as combustible. The National Fire Protection Association (NFPA) classification, which is most widely used, is further summarized in table 10-2. The Department of Transportation (DOT) classification system requires a red label to be placed on containers of materials with a flash point of 80°F or below — showing them to be hazardous.

Class	Flash Point, °F
1	below 100
2	100 to 139
3	140 and above

Note: Classes 1 and 3 are subdivided by taking boiling point into consideration. Class 2 is not subdivided.

Class	Flash Point, °F	Boiling Point, °F
1a	below 73	below 100
1b	below 73	100 and above
1c	73–99	
3a	140 to 199	
3b	200 and above	

Table 10-2. NFPA Classification System Based On Flash Point For Rating of Flammable Hazardous Materials.

IGNITION TEMPERATURE

Combustion is also defined in terms of ignition temperature, which has no relationship to the flash point of a material. The ignition temperature of any material — gas, liquid, or solid — is the temperature to which it must be heated to ignite. It is not necessary that the entire mass of material be heated to the ignition temperature for ignition to occur. A local "hot spot" is sufficient. Ignition occurs at the site which has reached the ignition temperature. The heat released by the combustion process propagates the flame to the entire material. Ignition temperature is a valuable qualitative concept, but because of its qualitative nature, it is not used as a basis for classification of hazardous materials.

Three primary conditions must be met for combustion to be initiated.

- The temperature must be above the flash point.

- The vapor-air composition must fall between the limits determined by the upper and lower flammable limits.

- The temperature must equal or exceed the ignition temperature.

WATER SOLUBILITY

The *water solubility* (ability to mix with water) of a material would seem to be an important factor. Water is a widely used and very effective extinguishing agent for most fires because it is an effective cooling agent. In a wood fire, water effectively cools the burning wood below its ignition temperature. Water is also effective against some low flash point materials (Class 2 or 3 liquids with flash points above 100°F) because, in principle, it cools them below the flash point. Other low flash point materials (Class 1 with flash points below 100°F) cannot be effectively extinguished with water, since water (hydrant) is nearly always between 50° and 60°F.

If an attempt is made to reduce the flash point of a typical water-soluble material such as acetone or alcohol by dilution with water, large quantities of water must be added. This is impractical for large fires. It is effective only for small fires or in isolated instances involving larger quantities where containment of the diluted materials is possible.

SPECIFIC GRAVITY

The *specific gravity* of a hazardous material is also of secondary importance. Although it is not directly related to fire initiation, it has a relationship to some fire fighting procedures. *Specific gravity* is defined as the ratio of the density of a material to that of water. Thus, if a material has a specific gravity of less than 1, it is lighter than water. Conversely, if it is heavier than water, it has a specific gravity greater than 1.

Materials such as gasoline and kerosene have densities less than 1. They float on water. This is why spraying a gasoline fire with water often spreads the fire. Materials such as oils and fats have densities greater than 1. Water can be effective against such materials if it can be spread across the surface (and if the material has a flash point higher than 100°F). Special care must be taken to avoid getting water below the surface of the oil or fat.

Volatilization of water will increase the fire hazard by atomizing the oil or fat.

VAPOR DENSITY

A knowledge of the vapor density of a material is useful in predicting whether a material is heavier or lighter than air. It is most valuable in avoiding potential hazardous conditions. A material such as ether is much heavier than air. It will flow across floors and down steps until it reaches an ignition source. Natural gas is just the opposite. Being lighter than air, it rises to the higher areas.

Vapor density is defined as the ratio of the density of a gas to an equal volume of air. It is similar to the specific gravity definition for a liquid. In a combustion situation, the significance of vapor density tends to be obscured because of heating and mixing processes. Even dense vapors can be carried upward into the fire zone.

The significance of many other physical properties, such as boiling point, melting point, vapor pressure, particle size of solids, and thermal decomposition points, could be discussed and debated. In general, they are significant only in a limited context. These factors will be discussed in conjunction with special situations as they are encountered, however.

REVIEW

1. The flash point of a hazardous material can be predicted from
   a. Boiling point.   c. Vapor pressure.
   b. Melting point.   d. Empirical formula.

2. If a hazardous material is present at a concentration level below the lower flammable limit it
   a. Will not burn well.   c. Is difficult to ignite.
   b. Will not burn at all.   d. Will burn vigorously.

3. If a hazardous material is present at a concentration level above the upper flammable limit it
   a. Will not burn well.   c. Is difficult to ignite.
   b. Will not burn at all.   d. Will burn vigorously.

4. A material presents the greatest flammable hazard when it is

   a. Below the lower flammable limit.
   b. Above the upper flammable limit.
   c. Between the upper and lower flammable limits.
   d. Depends on random conditions.

5. The upper flammable limit for gasoline and kerosene is most nearly

   a. 90% in air.   c. 45% in air.
   b. 75% in air.   d. 10% in air.

6. The most important physical property in determining the flammability hazard of gasoline is

   a. High flash point.   c. Ignition temperature.
   b. Low flash point.   d. High vapor density.

7. Whenever a gasoline-air mixture falls within the upper and lower flammable limits, the flammable hazard is

   a. Highest near the upper limit.   c. Highest near the mid-point.
   b. Highest near the lower limit.   d. Equal anywhere within the limits.

8. The flash point is always lowest for a

   a. Gas.   c. Solid.
   b. Liquid.   d. Plastic.

9. Which of the following has the highest flash point?

   a. Hydrogen.   c. Kerosene.
   b. Gasoline.   d. Coal.

10. A DOT red label on a container indicates

    a. A low flash point material.   c. A high flash point material.
    b. A medium flash point material. d. None of these.

11. A combustible liquid has a flash point

    a. below 80°F.   c. above 100°F.
    b. below 100°F.   d. All of these.

12. Which of the following materials has a specific gravity greater than 1?

    a. Wood.   c. Lubricating oil.
    b. Gasoline.   d. Kerosene.

Unit 11 Reactivity With Oxygen

Chemically, oxygen is a reactive element. It combines with all of the metallic elements except gold, platinum, and palladium. Some metals such as sodium, potassium, and calcium react rapidly with oxygen. They are often referred to as *pyrophoric* metals because they react spontaneously with air. Others such as magnesium react slowly at ambient conditions. Copper and mercury among others, react only at elevated temperatures.

Some of the more common reactions of metals with oxygen are shown in table 11-1. Combustion of metals can be hazardous both in terms of heat evolution and toxicity. This is discussed in depth in unit 15. For now, the intent is to develop a feeling for the essential similarity involved in the combustion of all metals.

Atomic Number	Reaction	Product
3	$4Li + O_2 \rightarrow 2Li_2O$	Lithium oxide
11	$4Na + O_2 \rightarrow 2Na_2O$	Sodium oxide
19	$4K + O_2 \rightarrow 2K_2O$	Potassium oxide
4	$2Be + O_2 \rightarrow 2BeO$	Beryllium oxide
12	$2Mg + O_2 \rightarrow 2MgO$	Magnesium oxide
20	$2Ca + O_2 \rightarrow 2CaO$	Calcium oxide
38	$2Sr + O_2 \rightarrow 2SrO$	Strontium oxide
22	$Ti + O_2 \rightarrow TiO_2$	Titanium dioxide
40	$Zr + O_2 \rightarrow ZrO_2$	Zirconium dioxide
26	$4Fe + 3O_2 \rightarrow 2Fe_2O_3$	Iron oxide
29	$2Cu + O_2 \rightarrow 2CuO$	Copper oxide
47	$2Ag + O_2 \rightarrow AgO$	Silver oxide
30	$2Zn + O_2 \rightarrow 2ZnO$	Zinc oxide
48	$2Cd + O_2 \rightarrow 2CdO$	Cadmium oxide
80	$2Hg + O_2 \rightarrow 2HgO$	Mercury oxide
13	$4Al + 3O_2 \rightarrow 2Al_2O_3$	Aluminum oxide

Table 11-1. Reactions of Metals With Oxygen.

Nonmetals also combine with oxygen at sufficiently high temperatures. They are generally nonreactive with oxygen at ambient conditions. The common reactions of nonmetals with oxygen are shown in table 11-2.

Atomic Number	Reaction	Product
1	$2H_2 + O_2 \rightarrow 2H_2O$	Hydrogen oxide (water)
5	$4B + 5O_2 \rightarrow 2B_2O_5$	Boric oxide
6	$C + O_2 \rightarrow CO_2$	Carbon dioxide
14	$Si + O_2 \rightarrow SiO_2$	Silicon dioxide (sand)
7	$2N + O_2 \rightarrow 2NO$	Nitric oxide
15	$4P + 5O_2 \rightarrow 2P_2O_5$	Phosphorus pentoxide
16	$S + O_2 \rightarrow SO_2$	Sulfur dioxide

Table 11-2. Reactions of Nonmetals With Oxygen.

Everyday experience indicates that some of the common fire extinguishing agents are water (H_2O), carbon dioxide (CO_2), and silicon dioxide (SiO_2). The fundamental reason for their use is that they have burned already. Therefore, they do not chemically react with oxygen and can effectively exclude oxygen from the combustion zone and remove heat by cooling. Water, carbon dioxide, and sand are selected from among the many choices because they are available in quantity and are relatively inexpensive.

Combustion, or burning, is simply the rapid combination of oxygen with a reactive material. It is possible to use the periodic table, shown in figure 1-3, on page 5, to predict the relative reactivity of metals with oxygen. The metals listed in Group I on the

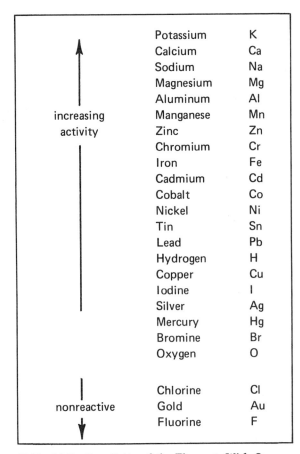

increasing activity	Potassium	K
	Calcium	Ca
	Sodium	Na
	Magnesium	Mg
	Aluminum	Al
	Manganese	Mn
	Zinc	Zn
	Chromium	Cr
	Iron	Fe
	Cadmium	Cd
	Cobalt	Co
	Nickel	Ni
	Tin	Sn
	Lead	Pb
	Hydrogen	H
	Copper	Cu
	Iodine	I
	Silver	Ag
	Mercury	Hg
	Bromine	Br
	Oxygen	O
nonreactive	Chlorine	Cl
	Gold	Au
	Fluorine	F

Table 11-3. Reactivity of the Elements With Oxygen.

left are more reactive than those to the right. Thus, lithium is more reactive than beryllium, and sodium is more reactive than magnesium. In any period of elements, those at the top are more reactive with oxygen than those at the bottom. In Group I, the elements in decreasing order of reactivity are hydrogen, lithium, sodium, and potassium. The relative order of reactivity of the more common elements is shown in table 11-3.

A wide variety of other materials are encountered in combustion and explosion situations. Many of them are partially oxidized substances such as carbon monoxide which can be oxidized to completion with the availability of additional oxygen.

$$2C + O_2 \rightarrow 2CO$$
$$2CO + 2O_2 \rightarrow 2CO_2$$

Additional typical reactions of this type are shown in table 11-4. The most familiar example of a partially oxidized material is that of wood. All other partially oxidized materials react similarly with oxygen.

Nitromethane	$4CH_3NO_2 + 3O_2 \rightarrow 4CO_2 + 2N_2 + 6H_2O$
Methyl alcohol	$2CH_3OH + 3O_2 \rightarrow 2CO_2 + 4H_2O$
Sulfur dioxide	$2SO_2 + O_2 \rightarrow 2SO_3$
Nitric oxide	$2NO + O_2 \rightarrow 2NO_2$
Wood	$C_6H_{10}O_5 + 6O_2 \rightarrow 6CO_2 + 5H_2O$

Table 11-4. Combustion of Some Typical Partially-Oxidized Materials.

REVIEW

1. Which of the following elements is combustible with air under normal fire conditions?

   a. Hydrogen.   c. Iron.
   b. Nickel.   d. Gold.

2. Which of the following elements is pyrophoric?

   a. Sodium.   c. Hydrogen.
   b. Magnesium.   d. Iron.

3. Magnesium fires are particularly hazardous because

   a. Magnesium is slightly reactive with oxygen.
   b. Magnesium is pyrophoric.
   c. Magnesium is exothermic.
   d. Magnesium is volatile.

4. Which of the following metals is most reactive with oxygen?

   a. Zinc.

   b. Aluminum.

   c. Magnesium.

   d. Cadmium.

5. Which of the following elements is a nonmetal?

   a. Lithium.

   b. Chlorine.

   c. Mercury.

   d. Calcium.

6. Which of the following nonmetals is most reactive with oxygen?

   a. Chlorine.

   b. Fluorine.

   c. Bromine.

   d. Iodine.

7. Which of the following nonmetals has been incompletely oxidized?

   a. CO_2.

   b. CO.

   c. SO_2.

   d. NO_2.

8. The most reactive metals in the Periodic Chart are those shown on the

   a. Left in Group 1.

   b. Right in Group 7.

   c. Middle in Group 3.

   d. Bottom in Group 1.

9. The most reactive nonmetals in the Periodic Chart are those shown on the

   a. Left in Group 1.

   b. Right in Group 7.

   c. Middle in Group 3.

   d. Bottom in Group 7.

10. Which of the following compounds is completely oxidized?

   a. Methyl alcohol.

   b. Nitromethane.

   c. Carbon monoxide.

   d. Silicon dioxide

Unit 12 Reactivity With Water

The principles involved in the discussion of the reactivity of water are a logical extension of the concepts developed in the preceding unit. It was seen that most metals and all nonmetals react with oxygen with the formation of oxides when the reaction is carried to completion. Water was one of the examples cited. Water is the end-product of the combustion of hydrogen and oxygen. It is an effective fire extinguishing agent for many types of fires. It must be remembered, however, that water is in turn reactive with many other substances.

When water is used — perhaps inadvertently — in a hazard situation involving water-reactive materials, an explosion can result, the fire may be intensified, or at the very least, the water is ineffective in extinguishment. Calcium carbide is an example of such a material. When water is applied to this grayish-black solid, acetylene gas is produced. This type of reaction results in explosive conditions.

The reactions of typical common metals with water are shown in table 12-1. The reactivity of these same metals with oxygen was shown in table 11-1. The nature of the chemical reactions are very similar in the sense that oxides are formed with oxygen and hydroxides are formed with water. Hydroxides are often referred to as hydrated oxides. A

Atomic Number	Reaction	Product
3	$2Li + 2H_2O \rightarrow 2LiOH + H_2$	Lithium hydroxide
11	$2Na + 2H_2O \rightarrow 2NaOH + H_2$	Sodium hydroxide
19	$2K + 2H_2O \rightarrow 2KOH + H_2$	Potassium hydroxide
4	$Be + 2H_2O \rightarrow Be(OH)_2 + H_2$	Beryllium hydroxide
12	$Mg + 2H_2O \rightarrow Mg(OH)_2 + H_2$	Magnesium hydroxide
20	$Ca + 2H_2O \rightarrow Ca(OH)_2 + H_2$	Calcium hydroxide
38	$Sr + 2H_2O \rightarrow Sr(OH)_2 + H_2$	Strontium hydroxide
22	$Ti + 4H_2O \rightarrow Ti(OH)_4 + 2H_2$	Titanium hydroxide
40	$Zr + H_2O \rightarrow Zr(OH)_4 + 2H_2$	Zirconium hydroxide
26	$2Fe + 6H_2O \rightarrow 2Fe(OH)_3 + 3H_2$	Iron hydroxide
29	$Cu + 2H_2O \rightarrow Cu(OH)_2 + H_2$	Copper hydroxide
47	$Ag + 2H_2O \rightarrow Ag(OH)_2 + H_2$	Silver hydroxide
30	$Zn + 2H_2O \rightarrow Zn(OH)_2 + H_2$	Zinc hydroxide
48	$Cd + 2H_2O \rightarrow Cd(OH)_2 + H_2$	Cadmium hydroxide
80	$Hg + 2H_2O \rightarrow Hg(OH)_2 + H_2$	Mercury hydroxide
13	$2Al + 6H_2O \rightarrow 2Al(OH)_3 + 3H_2$	Aluminum hydroxide

Table 12-1. Reactions of Metals With Water.

familiar example is the hydration of lime (calcium oxide) to form calcium hydroxide in the cement-making process. This sequence of oxidation and hydration reactions is shown.

$$2Ca + O_2 \rightarrow 2CaO$$
$$CaO + H_2O \rightarrow Ca(OH)_2$$

The direct reaction of a metal with water produces a hydroxide and hydrogen.

$$Ca + 2H_2O \rightarrow Ca(OH)_2 + H_2$$

The water-reactive metals are generally shipped in airtight containers or under an inert medium such as mineral oil or, in some cases, kerosene. In this manner, their rapid reaction with moisture is inhibited. Sodium, for example, reacts with water to give hydrogen and sodium hydroxide as a by-product. As this reaction takes place, the heat evolved is sufficient to ignite the hydrogen causing an explosion with the splattering of caustic. This is the reason that complete protective clothing should be worn when handling water reactive metal spills or accidents.

For metal fires, dry graphite, sodium chloride (table salt) or potassium chloride can be employed. Met-L-X, a commercial extinguishing agent, contains sodium chloride and a plastic agent which melts onto the metal thus covering completely different areas and shapes, thus excluding oxygen and cooling the metal below ignition temperature. Extinguishing agents, such as foam, carbon dioxide, and halogenated compounds, should never be used on metal fires. These only make conditions worse.

Several salient aspects of the hazards associated with combustion of metals can be seen by inspection of table 12-1, page 45.

- Hydrogen is evolved when a metal reacts with water. For that reason, water can not be used as an extinguishing agent for metal fires.

- All metals form hydroxides as reaction products with water. The most familiar example of a hydroxide is lye (sodium hydroxide). Reaction of a metal with water produces hazardous products which are caustic like lye. In addition to the fire hazard, corrosive combustion products are formed which are highly toxic.

- Although metal oxides are not combustible with oxygen, their presence in a fire area can be an indirect hazard. If large quantities of any of the metal oxides are deluged with water, heat is evolved and they are converted to caustic materials. This is particularly hazardous in the case of light metal oxides or hydroxides.

The same periodic law correlations discussed in unit 11 for the reaction of metals with oxygen apply for water, also. The relative reactivity of water with metals follows the same order presented for oxygen in table 11-3.

The metals to the left-hand side of the periodic chart are most reactive. Thus, the Group I elements (Li, Na, K) are much more reactive and hazardous with water than the Group II elements (Be, Mg, Ca, Sr). Similarly, lithium is more reactive and hazardous than sodium. This relative hazard rating is based on the rate at which hydrogen is evolved. The relative caustic strength of the hydroxides follows the same pattern: sodium hydroxide is more caustic than calcium hydroxide.

The reaction of several typical nonmetals is shown in table 12-2. The most common nonmetals which react directly with water are boron, phosphorus, and the halogens. *Halogen* is a term for the group of elements including fluorine, chlorine, bromine, and iodine.

The other common nonmetals are carbon, nitrogen, and sulfur. They do not tend to react directly with water under ordinary conditions. The first step in their reaction

Atomic Number	Reaction	Product
5	$B + H_2O \rightarrow H_3BO_3$	Boric acid
15	$P + H_2O \rightarrow H_3PO_4$	Phosphoric acid
9	$2F_2 + 2H_2O \rightarrow 4HF + O_2$	Hydrofluoric acid
17	$2Cl_2 + 2H_2O \rightarrow 4HCl + O_2$	Hydrochloric acid
35	$2Br_2 + 2H_2O \rightarrow 4HBr + O_2$	Hydrobromic acid
6	$C + O_2 \rightarrow CO_2 \xrightarrow{H_2O} H_2CO_3$	Carbonic acid
7	$N + O_2 \rightarrow NO \xrightarrow{H_2O} HNO_3$	Nitric acid
16	$S + O_2 \rightarrow SO_2 \xrightarrow{H_2O} H_2SO_4$	Sulfuric acid

Table 12-2. Reaction of Nonmetals With Water.

sequence is with oxygen. The oxides in turn react readily with water to form acids.

The primary hazards associated with the combustion of nonmetals can be seen by inspection of table 12-2.

- All nonmetals form acids by direct reaction (B, P, F_2, Cl_2, Br_2) or by indirect reaction of the oxides (C, N, S).

- All acids are extremely toxic. Like the corresponding hydroxides formed from metals, the acids are classified as corrosive hazardous materials.

- Although the toxicity hazard is relatively high for the nonmetals, the combustion hazard, relative to the metals, is low. Hydrogen is evolved when metals react with water, whereas this is not true for the nonmetals.

- Water is an effective extinguishing agent for all the nonmetals except boron, phosphorus, and the halogens. Fortunately, most fire situations involve the other nonmetals, carbon, nitrogen, and sulfur.

REVIEW

1. Most metals and all nonmetals react with

   a. Oxygen.
   b. Water.
   c. Carbon dioxide.
   d. Sodium bicarbonate.

2. The reaction product of a combustion reaction is

   a. Water.
   b. An oxide.
   c. A chloride.
   d. Any of these.

3. The reaction of metals with oxygen and water

   a. Are equally hazardous.
   b. Vary in the degree of hazard.
   c. Tend to be more hazardous with water.
   d. Tend to be more hazardous with oxygen.

4. The reaction of a metal with oxygen produces
    a. Hydrogen.   c. An oxide.
    b. Oxygen.   d. A hydroxide.

5. The reaction of a metal with water produces
    a. Hydrogen.   c. An oxide.
    b. Oxygen.   d. All of these.

6. The reaction of a metal and a metal oxide with water both produce
    a. Acid.   c. Oxide.
    b. Hydrogen.   d. Hydroxide.

7. Metal oxides are hazardous because
    a. They are highly flammable.
    b. They are explosives.
    c. They form caustic hydroxides with water.
    d. They are endothermic materials.

8. Sodium is more reactive than potassium with
    a. Oxygen.   c. Both oxygen and water.
    b. Water.   d. None of these.

9. Calcium is less reactive than magnesium with
    a. Oxygen.   c. Both oxygen and water.
    b. Water.   d. None of these.

10. Which of the following is a halogen compound?
    a. NaF.   c. Phosphorus.
    b. Boron.   d. Helium.

11. Nonmetals react with oxygen
    a. Never.   c. At high temperature.
    b. At ambient temperature.   d. Depending largely on pressure.

12. Nonmetals react with water to form
    a. Oxides.   c. Acids.
    b. Hydroxides.   d. Bases.

13. Water is an effective extinguishing agent for
    a. All metals.   c. All nonmetals.
    b. Some nonmetals.   d. None of these.

14. Nonmetals occur in the periodic table only in
    a. Group I.   c. Group III.
    b. Group II.   d. Group IV.

15. Which of the pairs of chemical names refer to the same compound?
    a. Hydrofluoric acid and hydrochloric acid.
    b. Nitric acid and nitric oxide.
    c. Carbonate and carbonic acid.
    d. Hydrofluoric acid and hydrogen fluoride.

Unit 13 Reactivity With Chemical Extinguishing Agents

Fire extinguishers are classified on the basis of their effectiveness in extinguishing various types and sizes of fires. Fires are classified into four basic types: Classes A, B, C, and D.

CLASS A FIRES

Class A fires involve combustion of ordinary materials, such as wood and paper. Class A extinguishers utilize the combined effects of quenching and cooling by water. The simplest Class A extinguisher is a water vessel in which the operator expels the water with a manual pump. Other versions utilize a gas pressurization cartridge to automatically expel the water or a tank under air pressure. The soda-acid extinguisher involves a chemical reaction to expel water; this extinguisher, now outdated and no longer in "common" use, contained two solutions: an acid and a carbonate solution. The two solutions were mixed to activate the extinguisher. The carbon dioxide formed by the reaction of the acid and carbonate expelled water through pressure buildup.

$$H_2SO_4 + Na_2CO_3 \rightarrow Na_2SO_4 + CO_2 + H_2O$$

This was just a convenient way of expelling the water stream. The effective extinguishing agent was water. The problem with this extinguisher was that the sulfuric acid absorbed the water over a period of time and the extinguisher became deactivated.

CLASS B FIRES

Class B fires involve combustion of the more flammable materials such as liquids and greases. Class B extinguishers operate by means other than cooling. They include foam, dry chemical, halocarbon, and inert gas extinguishing agents.

- The foam type extinguishers spread a layer of durable foam over the burning material and exclude oxygen.

- The dry chemical extinguishers utilize a variety of methods for excluding oxygen from the fire.

  a. The carbon dioxide extinguisher expels carbon dioxide to smother the flame.

  b. The dry chemical extinguisher smothers the flame with sodium or potassium bicarbonate, or monoammonium phosphate.

  c. The halocarbon extinguishers suffocate the flame with bromine and chlorine-containing liquids. These halogens also cool the flame by vaporization, and act as free radical quenchants.

CLASS C FIRES

Class C identifies fires in live electrical systems. The primary requirement in controlling a Class C fire involving live electrical systems is the use of an electrically nonconductive extinguishing agent. Water is clearly not suitable. Any of the Class B extinguishing agents listed in table 13-1 are suitable because they are not electrically conductive. For this reason, Class B extinguishers are also suitable for Class C fires.

CLASS D FIRES

Class D refers to metal fires. Additional types of dry powder extinguishing materials used in special situations are summarized in table 13-1. These special situations involve extinguishment of metal or metallic alloy fires, such as sodium, magnesium, titanium, potassium, and lithium.

- Fine sand (SiO_2) is very effective for small metal fires. It is chemically inert and has a high heat capacity, which is helpful in cooling the reaction zone. It functions primarily by excluding oxygen. Obviously, the sand must be dry since metals are reactive with water.

- Finely-divided graphite powder (C) is used in the same way.

- Mixtures of salt (NaCl) and a polymeric binder are very effective for suffocating metal fires. This type of Class D extinguisher is sold as Met-L-X. When applied to the fire zone, the plastic melts, sealing the fire off in a thick layer of salt.

- The use of inert gases (helium, argon, nitrogen) can be applied only in special circumstances. If the metal fire is confined in a limited volume, displacement of oxygen with an inert gas may be effective. However, a metal fire is often in an open area or too large to be effectively controlled with an inert gas.

FIRE SIZE

Chemical fire extinguishers are also rated in terms of the maximum size of the fire that the extinguisher can control. Rating of both chemical type and size provides a good basis for selecting the proper extinguisher for a given fire hazard situation.

Class A extinguishers are rated according to their ability to extinguish standard wood fires of various sizes. A rating of Class 6-A indicates a threefold higher extinguishing capacity than that of a Class 2-A extinguisher. The Class B rating system is similar except that it is based on a standard gasoline fire. A

Dry Chemical — For Flammable Liquids (Class B)	
Carbon dioxide	CO_2
Sodium bicarbonate	$NaHCO_3$
Potassium bicarbonate	$KHCO_3$
Halogenated	$CBrF_3$
Inert Gas — For Combustible Metals	
Helium	He
Argon	Ar
Nitrogen	N_2
Dry Powder — For Combustible Metals	
Sodium Chloride	NaCl
Sand	SiO_2
Graphite	C

Table 13-1. Chemical Fire Extinguishing Agents.

Class 8-B rating indicates a twofold higher extinguishing capacity than that of a Class 4-B extinguisher.

PROPERTIES OF EXTINGUISHING AGENTS

The various extinguishing agents suitable for use in Classes A, B, C and D fires were chosen for their chemical properties. Class D fires are metal fires by definition. The dry powders listed in table 13-1 are selected from the numerous candidates on the basis of their ability to extinguish metal fires. Sodium chloride (NaCl) is a good extinguishing agent because it is not reactive, has a relatively high melting point, and is cheap. In the case of a sodium fire, sodium chloride is an effective extinguishing agent because it is inert to both oxygen and sodium.

$$O_2 + NaCl \rightarrow \text{no reaction}$$
$$Na + NaCl \rightarrow \text{no reaction}$$

The combustion reaction which the sodium chloride prevents is the formation of sodium oxide.

$$4Na + O_2 \rightarrow 2Na_2O$$

The danger caused by the reaction of sodium oxide with water to form caustic materials was pointed out earlier.

Sodium chloride is effective against other metal fires also. In the case of a magnesium (Mg) fire, the combustion reaction which must be prevented is the formation of magnesium oxide.

$$O_2 + 2Mg \rightarrow 2MgO$$

To be effective against a magnesium fire, sodium chloride must be nonreactive with oxygen and sodium chloride.

$$O_2 + NaCl \rightarrow \text{no reaction}$$
$$Mg + NaCl \rightarrow \text{no reaction}$$

The first equation simply indicates that salt (sodium chloride) does not burn in air. This statement is easy to accept because it is known to be true from everyday experience. The student is justified in wondering why the equation is not written as

$$O_2 + 4NaCl \rightarrow 2Na_2O + 2Cl_2$$

The reason is that chlorine (Cl_2) is a much stronger oxidizing agent than oxygen (O_2); therefore, it has an even greater driving force to react with Na_2O. As a matter of fact, this particular reaction would proceed with explosive violence.

$$O_2 + 4NaCl \rightarrow 2Na_2O + 2Cl_2$$
$$2Na_2O + 2Cl_2 \rightarrow 4NaCl + O_2$$

The net effect is expressed by the chemical equation

$$O_2 + NaCl \rightarrow \text{no reaction}$$

The second equation may seem strange to the fire science student, but it is gained from everyday experience by the laboratory chemist.

$$Mg + NaCl \rightarrow \text{no reaction}$$

The chemical logic is very similar to that just discussed for the reaction of sodium chloride and oxygen. *If* the reaction were to occur, sodium would be liberated.

$$Mg + 2NaCl \rightarrow MgCl_2 + 2Na$$

Sodium is a more reactive metal than magnesium; therefore, it would displace magnesium from magnesium chloride.

$$2Na + MgCl_2 \rightarrow Mg + 2NaCl$$

Many inert salts other than sodium chloride could be chosen for use. Some of them would be slightly better, but salt is by far the cheapest material of its type available. For that reason, it is most widely used. Graphite (C) and sand (SiO_2) function in a similar manner. An analysis of these systems is used as a student exercise in the review section.

The Class B extinguishers (carbon dioxide, sodium bicarbonate, potassium bicarbonate, potassium chloride and monoammonium phosphate) are not suitable for use in Class D fires involving metals. If carbon dioxide were used in an attempt to extinguish a magnesium fire, it would fail as a result of the following sequence of reactions:

$$Mg + CO_2 \rightarrow MgO + CO + \text{heat}$$
$$Mg + CO \rightarrow MgO + C + \text{heat}$$
$$C + O_2 \rightarrow CO_2$$

All carbonates react in a manner similar to carbon dioxide and are, therefore, unsuitable for metal fires. On the other hand, the carbonates and carbon dioxide are suitable for Class B fires involving flammable liquids.

Fire fighters have long been familiar with carbon tetrachloride as an extinguishing agent. However, because of its reaction and formation of toxic vapors, it has been discarded as an extinguisher. Other, safer forms have taken its place.

The halogens (chlorine, fluorine, bromine and iodine) are very effective in reacting with the products of combustion. The halogens have the ability to "quench" these products and thus "cool" the reaction and limit its extent. They are synthesized with carbon and so are called halocarbons.

Halocarbons "break down" very easily. It is this breaking down into free radicals (fragments of a molecule) that extinguishes fires. For instance, the reaction of chlorobromomethane (CH_2BrCl) with natural gas (methane) proceeds with the chlorobromomethane producing a free radical of chloromethyl and bromine. These free radicals are highly active and the bromine radical reacts with methane to form a methyl radical plus the acid HBr. The acid in turn reacts with the hydroxyl radical (formed constantly in any fire) to produce water and again the bromine radical. This reaction continues rapidly until all the fuel is used up or the ignition temperature is lowered.

$$CH_2BrCl \qquad + \text{ heat} \rightarrow CH_2\overset{\cdot}{C}l + Br$$
$$\text{Methane } (CH_4) \quad + \overset{\cdot}{Br} \rightarrow \overset{\cdot}{C}H_3 + HBr$$
$$\overset{\cdot}{O}H \qquad + HBr \rightarrow \overset{\cdot}{Br} + H_2O$$
$$\text{Final reaction: } CH_4 + \overset{\cdot}{O}H \rightarrow \overset{\cdot}{C}H_3 + H_2O$$

The order of the effectiveness of these halocarbons is listed: iodine > bromine > chlorine > fluorine. Iodine, however, is not used because compounds such as CH_2ICl are not stable enough for field use. They spontaneously decompose slowly, even in an unopened container.

Some of the common halocarbon extinguishing agents are shown.

Name	Formula	Commercial Name
trifluorobromo-methane	$CBrF_3$	Halon 1301
difluorochloro-bromomethane	$CBrClF_2$	Halon 1211
difluorodibromo-methane	CBr_2F_2	Halon 1202
difluorobromo-ethylene	$CBrF_2CBrF_2$	Halon 2402

REVIEW

1. What is the classification of a fire involving the combustion of a wooden warehouse in which large quantities of sodium are stored?

    a. Class A.
    b. Class B.
    c. Class C.
    d. Class D.

2. What is the classification of a fire involving the combustion of a wooden warehouse in which large quantities of electrical supplies are stored?

    a. Class A.
    b. Class B.
    c. Class C.
    d. Class D.

3. Which of the following extinguishing agents is suitable for extinguishing a calcium fire?

    a. Carbon tetrachloride.
    b. Sodium bicarbonate.
    c. Carbon dioxide.
    d. Sodium chloride.

4. Magnesium reacts with

    a. Water.
    b. Carbon dioxide.
    c. Sodium bicarbonate.
    d. All of these.

5. Which of the following is an effective choice for Class C fires?

    a. SiO_2.
    b. Graphite.
    c. Sodium chloride.
    d. All of these.

6. Which of the following is suitable for Class D fires?

    a. Sodium chloride   c. Graphite
    b. Sand   d. All of these

7. Which of the following chemical reactions properly describes the course of an ordinary fire situation involving the combustion of methane?

    a. $2CH_4 + 3O_2 \rightarrow 2CO + 4H_2O$ c. $CH_4 + O_2 \rightarrow CO_2 + 2H_2$
    b. $C_2H_6 + 2O_2 \rightarrow 2CO_2 + 3H_2$ d. $CH_4 + 2O_2 \rightarrow CO_2 + 2H_2O$

8. Which of the following extinguishing agents is the most effective against a methane-air fire?

    a. Water   c. Helium
    b. Sand   d. Sodium bicarbonate

9. Graphite (C) is an effective extinguishing agent against a small magnesium fire. Which of the following chemical equations properly describes the reason for its effectiveness?

    a. $2Mg + C \rightarrow Mg_2C$ c. $Mg + O \rightarrow MgO$
    b. $Mg + C \rightarrow$ no reaction   d. $C + O_2 \rightarrow CO_2$

10. Sand (SiO_2) is effective against small magnesium fires. Which of the following explains its effectiveness?

    a. It is inexpensive.   c. Both a and b.
    b. It is chemically inert.   d. Neither a nor b.

11. Which of the following reactions properly describes the extinguishing characteristics of sand in a metal fire?

    a. $SiO_2 + Mg \rightarrow$ no reaction   c. $2SiO_2 + O_2 \rightarrow 2SiO_3$
    b. $SiO_2 + 2Mg \rightarrow 2MgO + Si$ d. None of these

12. Which of the following characteristics detracted from the desirability of carbon tetrachloride as a Class B extinguisher?

    a. It is expensive.   c. Its by-products are toxic.
    b. It is a liquid.   d. It contains carbon.

13. If potassium chloride were used instead of sodium chloride for extinguishing a metal fire, what would be the main disadvantage?

    a. It is a poorer quenchant than Na. c. It is toxic.
    b. It is not commercially available. d. It is more expensive.

14. Which of the following extinguishing agents would be most suitable against a Class C electrical fire?

    a. Class A extinguisher   c. Water
    b. Class B extinguisher   d. None of these

15. What classification would be assigned to a magnesium fire?

    a. Class A   c. Class C
    b. Class B   d. Class D

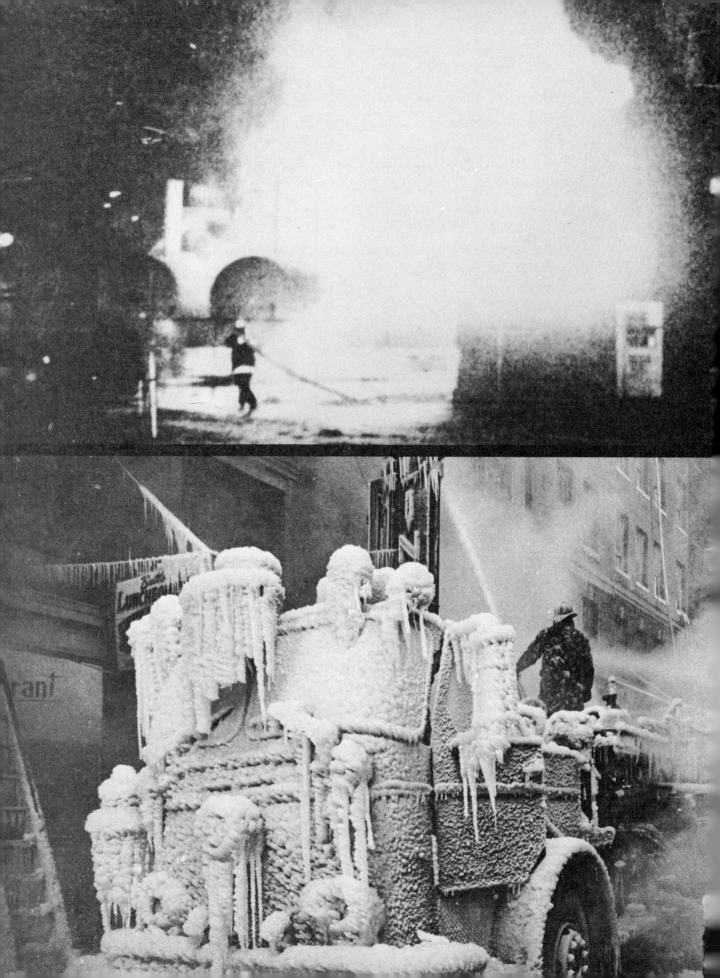

GASES

Unit 14 Compressed Gases

Gases make up one of the three states of matter. They are characterized by low density and ability to move freely. Compared to liquids and solids, they expand and contract easily with changes in pressure and temperature. In this unit, the common types of gases, pressurized and liquefied, are discussed. Gases offer a wide variety of uses: from medical applications to industrial processes to firefighting operations.

A *compressed gas* (including liquefied gases) is defined as any material or mixture that when enclosed in a container has an absolute pressure exceeding 40 psi at 70°F, or if the gas is flammable, the figure is 25 psig at 100°F. The family of compressed gases is large; air, liquefied petroleum gases, oxygen, nitrogen, and acetylene are some examples. Some of these gases are flammable, some nonflammable, some toxic, and others nontoxic. One has only to think of the materials around him that have their origin as compressed gases or are compressed gases to realize the wide uses and everyday occurrences of these gases.

Compressed gases may be divided into two areas: pressurized and liquefied. *Pressurized gases* are those that when compressed in a cylinder, do not liquefy because their boiling points are low (–150°F and lower). Continued depression of the boiling point introduces the area of cryogenics. Cryogenic gases are discussed later in unit 15. *Liquefied gases* exist in a cylinder in a liquid-vapor relationship. These gases have a boiling point closer to common temperatures from 32°F to –150°F. Butane, with a boiling point of 32°F,

is an example of a liquefied gas that does not give off vapors (boil) when the ambient temperature approaches freezing and thus is not useful as a fuel. This is why northern areas of the United States do not use butane systems. Propane, another example, has a boiling point of –45°F, and thus is useful in all areas. Chlorine, propylene, and ammonia (anhydrous) and refrigerant gases are other examples of liquefied gases. A gas which does not really fit into the two above categories is acetylene, which is a gas dissolved in a liquid.

Flammable gases, as well as nonflammable gases, are included in the list of containers or cylinders that are shipped and stored throughout the United States. All these gases are potential hazards if involved in fire conditions. They are protected from pressure buildup by certain safety devices built into their containers. Some, however, such as poisonous gases, do not have these devices.

SAFETY-RELIEF DEVICES FOR GAS CYLINDERS

The majority of gas cylinders in use today are equipped with some type of safety-relief device. The major function of this device(s) is to prevent rupture of the cylinder from the sudden increase in pressure that results from heat buildup. If the pressure can be released to the atmosphere, violent rupture is prevented.

Three types of devices generally used to protect cylinders today are (1) safety-relief valves, (2) frangible discs, and (3) fusible plugs. All three have the same primary function of relieving gases, but the mode of operation of each is different.

Safety-Relief Valve

A safety-relief valve generally is a part of the cylinder and is normally held in a closed position by spring force. The force holding the valve closed is dependent on the type of gas contained in the cylinder. Safety-relief valves generally reset at a pressure that is not less than the pressure in a normally charged cylinder at a temperature of 150°F. Increased pressure from the gas present will open the valve thus releasing the contents until the pressure is back to the safety limits. The valve then resets itself until another pressure buildup is encountered.

Frangible Disc

The frangible disc is different from the safety-relief valve mentioned earlier in that it cannot be reset. The metal disc is designed to burst at a predetermined pressure, thus releasing the entire contents of the cylinder. The bursting pressure of most frangible discs is designed to fail before the minimum test pressure of the cylinder is met. Frangible discs are found in most cylinders, used alone or in combination with safety-relief valves.

Fusible Plug

A third type of relief assembly is used for the more "active" gases. A fusible plug does not operate on pressure as the above valves do, but rather is temperature activated. Some gases, such as acetylene, decompose or polymerize when exposed to the elevated temperatures necessary to make pressure-relief devices activate. This results in rupture of the tank. This shrapnel type of explosion causes more damage than probably would occur if the gas inside the cylinder were allowed to escape in a controlled manner. The fusible plug is constructed of an alloy that melts at low temperatures to permit the compressed gas to escape rather than build up

in pressure. Acetylene tanks are so equipped with fusible plugs designed to function around 212°F. Another type of fusible plug operates at 165°F.

Compressed gas cylinders utilize a combination of frangible, safety-relief valves and fusible plugs. This combination allows for a wider range of safety. Not all gas cylinders are equipped with safety valves. Certain compressed gases, because of their poisonous nature, are not so equipped. Fluorine and the Class A poisons are examples.

TRANSPORTATION METHODS

Gases are shipped in a variety of ways — by rail, on the highway, on waterways, and in the air. These common means of transportation are discussed in the following paragraphs and are summarized in table 14-1.

Rail Shipping

Gases can be shipped by rail in tank cars or on TMU cars. Rail tank cars (single unit) are similar to the familiar oil tank cars. They are large pressure tanks that are part of the rail car. The maximum pressure allowed is 600 psig for these. Capacity ranges up to 10,000 gallons (measured in gallons of water). Larger tank cars have been built that carry up to 60,000 gallon capacity. These are used for liquefied petroleum products. The majority of the larger size in use have approximately a 30,000 gallon capacity. The tank cars are built either insulated or noninsulated, depending on the needs of the user.

The TMU (ton multi unit, or ton containers) rail car is easily distinguishable. It consists of a flat bed with fifteen cylinder tanks, each with a capacity of 100 to 200 gallons, depending on the gas carried. Pressures range from 500 to 1000 psig. Certain containers storing liquefied gases have two valves, one for vapor and one for liquid. These valves are covered by a hood which protects

GAS	RAILWAY	HIGHWAY	WATER	AIR (CARGO ONLY)
Acetylene	Cylinder	Cylinder	Cylinder	Cylinder
Air	Cylinder	Cylinder	Cylinder	Cylinder
Ammonia (anhy)	Cylinder, portable tank, TMU cars, tank cars	Cylinder, portable tank, TMU, tank	Tank	Cylinder
Argon	Cylinder, tank	Cylinder	Cylinder, tank	Cylinder
Boron trifluoride	Cylinder	Cylinder, tube trailer	Cylinder	Cylinder
Butadiene	Cylinder, tank, TMU	Cylinder, tank, portable tank	Portable tank, cylinder	
Carbon monoxide	Cylinder	Cylinder	Cylinder	Cylinder
Chlorine	Cylinder, tank, TMU	Cylinder, TMU	Cylinder, tank, TMU	Cylinder
Cyclopropane	Cylinder	Cylinder	Cylinder	Cylinder
Ether (dimethyl)	Cylinder, tank, TMU	Cylinder, TMU	Cylinder, tank	Cylinder
Ethane	Cylinder	Cylinder	Cylinder	Cylinder
Ethylene	Cylinder	Cylinder, tube trailer	Cylinder	Cylinder
Fluorine	Cylinder	Cylinder, tank (in special cases)	Cylinder	None
Refrigerants — fluorocarbons	Cylinder, tank, TMU	Cylinder, cargo tank, portable tank	Cylinder, TMU portable tank	NA
Helium	Cylinder, tank	Cylinder, tube trailer	Cylinder, tank	Cylinder
Hydrogen	Cylinder, tank	Cylinder, tube trailer	Cylinder	Cylinder
Hydrogen chloride	Cylinder, insulated tank	Cylinder, insulated tank, tube trailer	Cylinder	Cylinder
Hydrogen cyanide (Prussic acid)	Cylinder, tank cars	Cylinder	NA	NA
Hydrogen sulfide	Cylinder, TMU	Cylinder, TMU	Cylinder, tank	Cylinder
LP Gases: Butane, Isobutane Propane, Propylene Butene Isobutene	Cylinder, insulated and uninsulated tank car *also transported via pipeline	Cylinder, portable tank, cargo tank, semi-full trailer	Cylinder, portable tank, tank car	Cylinder
Methane	Cylinder *majority transported via pipeline	Cylinder	Cylinder	Cylinder
Methyl amines	Cylinder, insulated tank	Cylinder, tank	Cylinder, portable tank	Cylinder
Nitrogen	Cylinder, tank	Cylinder, tube trailer	Cylinder, tank	Cylinder
Nitrous oxide (laughing gas)	Cylinder, portable tank	Cylinder, portable tank, cargo tank	Cylinder	Cylinder
Oxygen	Cylinder, tank	Cylinder, tube trailer	Cylinder, tank car	Cylinder
Phosgene	Cylinder, TMU	Cylinder, TMU	Cylinder, TMU	None
Sulfur dioxide	Cylinder, tank car, TMU	Cylinder, cargo tank, portable tank, TMU	Cylinder, tank	Cylinder
Vinyl chloride	Cylinder, tank, TMU	Cylinder, tank, TMU	Cylinder, tank	Cylinder (inhibited only)
Vinyl methyl ether (inhibited)	Cylinder, tank, TMU	Cylinder, TMU	Cylinder, tank	Cylinder
-ethylamine -dimethylamine -trimethylamine	TMU		Portable tank	
Methyl chloride	Cylinder, TMU, tank	Cylinder, TMU, portable tank	Cylinder, portable tank	Cylinder
Methyl mercaptan	Cylinder, tank, TMU	Cylinder, portable tank, TMU	Cylinder, portable tank	Cylinder

SUT = single unit tank
TMU = ton multi unit

Table 14-1. Common Methods of Transporting Compressed Gases.

them against damage. Some TMU units contain fusible plugs, located in each head, which melt between 155°-165°F.

Highway Shipping

Three main types of vehicles serve the compressed gas industry as carriers: cargo tanks, portable tanks, and tube trailers.

The *cargo tanks* are similar to the permanently-mounted single unit tanks found on rail systems. The tanks are either mounted on small truck bodies or on larger beds in the form of semitrailers. The types of cargo tanks vary. Some are insulated, while others may have cooling coils or heating units. The type of cargo tank naturally depends on the gas being used. The most familiar are the LP gas trucks, which are insulated.

The *portable tanks* are cylindrical tanks with generally two flat legs on either end. These tanks closely resemble the larger cylindrical tanks found in LP gas shortage areas. They range from small units (120 gallon capacity) to larger capacity tanks (2,000 gallons or more). The pressures range from 100 psig to 500 psig.

Efficient utilization of gases can be achieved through the use of *tube trailers,* semitrailers carrying a series of large "tubes" or cylinders all joined together at a common head. These tube trailers usually carry oxygen, nitrogen, and helium (all pressurized, not liquefied). The use of tube cylinders allows for pressures up to 2,000 psig to be realized. The trailers are generally left at the user's location.

Air Shipping

Requirements are stringent for air shipping. Generally, only certain cylinders are allowed for air shipment. These cylinders are small; large quantities are not shipped in this manner.

STORAGE AREAS FOR GASES

Gases are stored in a variety of ways. For usage requiring only small amounts, the familiar compressed tanks are the most prevalent; for even smaller applications, "lecture bottles" of gas are frequently employed. These are small cylinders suitable for school or college demonstrations.

Use of larger quantities demands greater storing capacity. TMU units, tube trailers, or large portable tanks are used, depending upon the application and type of gas. The fire fighter should be aware of the existence and location of such tanks and their contents in his district. Preplanning is of extreme importance for such fixed locations.

MEDICAL GASES

Medical gases differ from the regular family of compressed gases only in their degree of purity. Some of the compressed gases can tolerate impurities to a small extent since they are used for industrial application. Gases used for medical applications, however, must be extremely high in purity — the impurities measured in parts per million instead of the regular parts per hundred (percent) commonly used. The gases used may be either liquefied or nonliquefied in nature, table 14-2.

Examples of the nonliquefied systems used are oxygen and helium. Mixtures may also be found, including carbon dioxide-oxygen and helium-oxygen.

Some medical gases are also found in the liquefied state. Carbon dioxide, cyclopropane, ethylene, nitrous oxide, and oxygen are examples. Oxygen is found both in the liquid and nonliquid state in hospitals.

The use of oxygen is obvious; its life-sustaining qualities make its use inevitable. However, some of the other gases are not so familiar.

Name	Formula	Flammable	Liq/Nonliq (Full Cylinder)	Specific Gravity	Flammable Limits Lower/Higher
Carbon Dioxide	CO_2	No	Liq below 88°F	1.53	- - -
Cyclopropane	C_3H_6	Yes	Liq	1.48	2.4 - 10.3
Ethylene	C_2H_4	Yes	Liq below 50°F Vapor-50°F higher	0.97	3.0 - 28.6
Helium	He	No	Vapor	0.14	
Nitrous Oxide	N_2O	No	Liq below 98°F Vapor-98°F higher	1.53	
Oxygen	O_2	No	Both liq and nonliquid	1.11	

Table 14-2. Medical Gases and Their Characteristics.

Cylinders used for medical applications, most notably the smaller portable cylinders, are color-coded to provide an easy means of preliminary identification. The colors are listed in table 14-3.

These colors must not be relied on for positive identification, however, since not all areas have adopted such color coding. The most positive identification is made by reading the label. When combinations of colors are used, the two colors are representative of a mixture of two gases. For example, for a carbon dioxide-oxygen mixture, the colors are gray (carbon dioxide) and green (oxygen).

Gas	Color
Carbon Dioxide	Gray
Carbon Dioxide-Oxygen	Gray-Green
Cyclopropane	Orange or Chrome-plated
Ethylene	Red
Helium	Brown
Helium-Oxygen	Brown-Green
Nitrous Oxide	Light Blue
Oxygen	Green

Table 14-3. Identification of Gases by Color of Cylinder.

OXYGEN: REPRESENTATIVE OF THE PRESSURIZED GASES

Most fire fighters are familiar with the pressurized gases. Oxygen is a representative example. Oxygen exists as a colorless, odorless, and tasteless gas. Found in air at a volume of 21 percent, it is categorized as nonflammable. It will not burn; however, it actively supports combustion. It is the support of combustion that causes the problems associated with oxygen. Many combustible materials when subjected to a 100 percent atmosphere of oxygen burn furiously. Even glowing steel wool burns intensely when subjected to pure oxygen.

Oxygen is used to increase the efficiency of blast furnaces, such as are used for purifying iron and copper. It is also used with acetylene tanks. The use of oxygen with acetylene increases flame temperature to 5720°F; this temperature makes the combination of oxygen and acetylene a versatile tool for rescue work.

Oxygen is unsurpassed in the medical field. It is used for resuscitation, heart stimulation, therapy purposes, and to sustain life in high and low pressure areas.

Shipping Oxygen

Oxygen is shipped in cylinders as a high-pressure gas. These cylinders have nominal capacities up to 330 cubic feet and generally are filled to 1800-2400 psig, depending on the cylinder size. As with all pressurized gases, the pressure gauge indicates the amount of pressurized gas remaining. This is entirely different from liquefied gases, where pressure always remains constant.

Oxygen is shipped on truck trailers which are fitted with long tubes, all joined together at a common head. These trailers also act as storage tanks for the user. The full trailers are left at the user's plant, and the empty ones are removed and refilled. The capacity of these tube trailers ranges up to 40,000 cubic feet of oxygen.

Tank cars transporting oxygen are similar to the previously mentioned tube trailers. The majority of pressurized gases is similar in high-pressure shipping and storage.

NATURAL GAS (METHANE)

The simplest of organic molecules, methane, is perhaps the most widely-used gas today. Methane is the major constituent in natural gas, is odorless, and is classified as a nontoxic gas.

Methane differs slightly from its companions, propane and butane, in that a molecule contains only one carbon atom compared to three in propane and four in butane. This slight difference, however, is enough to cause different behavior on the fire ground.

Natural gas contains small amounts of ethane and propane as constituents. In addition, an odoriferous chemical (called a mercaptan) is added in extremely small amounts (parts per million) to give natural gas its familiar odor.

Natural gas is widely used throughout the United States by industry and domestic users. Although it is extremely flammable,

extensive use of methane has brought a surprisingly small number of catastrophies. One that many will remember, perhaps, is the Brighton fire in Brighton, New York in the 1950s. A possible leak and subsequent explosion in a distribution vault opened pressure-regulating valves so that a thirty-pound pressure of gas entered into houses instead of the normal one-half pound. The sudden surge of gas resulted in torchlike flames coming out of gas water heaters, stoves, and furnaces in a twenty-five block area. Loss of life and destruction of homes resulted from the explosions and fires.

Another case occurred in New York City in January, 1967 when a large gas main ruptured, causing a thirteen-alarm fire in the Jamaica section. The fire, caused by a loose iron cover on a moisture collector, was fed from a twenty-four-inch gas main exerting 11.5 psig. Incoming apparatus, arriving at the scene before the fire started, stalled as it approached. The concentration of natural gas was so great that there was not sufficient oxygen for combustion for the engines.

Shortly after the apparatus arrived, ignition took place from an unknown source. Flames were reported to be 150 feet high and 20 feet wide. The ensuing radiant heat was so great that houses on both sides of the street began to burn.

Over 400 fire fighters and 70 pieces of apparatus fought the blaze as it burned an area of slightly more than two blocks. Fire fighters contained the blaze with master streams while gas company officials attempted to locate and turn off the gas main. Lost in the inferno were apparatus, automobiles, and the two blocks of homes. No deaths occurred despite the fact that the leak was first detected at 5:30 A.M.

Transporting Methane

Natural gas is piped under different pressures – low, medium, and high. In general,

IGNITED		LEAKING	
Outdoors		**Outdoors**	
A.	Notify gas company	A.	Notify gas company
B.	Let gas burn	B.	Evacuate area
C.	Evacuate area	C.	Eliminate ignition sources
D.	Protect surrounding exposures	D.	Shut off supply or plug
E.	Stand by at safe distance	E.	Stand by at safe distance
Indoors		**Indoors**	
A.	Notify gas company	A.	Notify gas company
B.	Shut supply off, if possible	B.	Evacuate building
C.	Extinguish and/or protect adjacent area until gas supply is off	C.	Shut off supply or plug
		D.	Eliminate sources of ignition
		E.	Ventilate (gas lighter than air)
		F.	Stand by at safe distance

Table 14-4. Suggested Natural Gas Procedures.

new homes and developments are serviced by the lower pressures, generally considered to be about 1/2 psi. Medium pressure lines are considered to have up to 60 psi, although this varies with different gas companies. High pressure lines are generally 60 psi or greater, and this too may vary with different gas companies. The higher pressure lines serve industries where gas consumption is high.

PROCEDURES FOR HANDLING GAS EMERGENCIES

The procedure for handling gas emergencies varies depending on whether or not fire is present. The local gas company is most helpful in supplying information on how to treat gas leaks in a particular district.

Table 14-4 shows suggested procedures for dealing with both ignited and unignited gases. In the event that fire has not occurred, an important point to remember is that it can at any instant. Fire fighters must turn the gas supply off and prepare for a fire or explosion while evacuation procedures are in progress.

While evacuation is taking place, a fundamental point to remember is that doorbells and light switches should be left alone!

The proper positioning of vehicles is also important. Too many times, while gas leaks in a building are being investigated, apparatus is parked directly in front of the building. Although no rule of thumb is available for proper positioning, the vehicle should be away from the building, in such a manner that fire or explosion will not interfere with its primary function.

Broken pipes or gas mains which cannot be plugged or turned off must have the leak controlled until gas company officials can turn the supply off. Broken mains which are on fire require protecting the exposures and controlling the fire. The direct flame coming from a broken main or pipe should be extinguished by shutting off the supply of gas and not by extinguishing agents. Extinguishment of the flame before the supply of gas is turned off may result in reignition of the vapors — possibly causing an explosion.

ACETYLENE

One of the most reactive of the organic series of gases is acetylene. Its reactivity stands out as highly dangerous with a NFPA rating of four.

Because of the three bonds located between the carbons, acetylene is highly reactive and highly flammable.

$$H - C \equiv C - H$$

Because of these factors, acetylene is differentiated from all other gases and is given special mention and special cylinders.

Acetylene is produced through the decomposition of calcium carbide by water.

$$CaC_2 + 2HOH \rightarrow Ca(OH)_2 + HC \equiv CH$$

This method, used years ago as a means of providing light for miners' helmets, is probably the most favorable. There are other methods used for acetylene manufacture — such as cracking of hydrocarbons or decomposition of methane — but the carbide offers economy in acetylene manufacture.

Welding and cutting applications may commonly be considered the major use of acetylene. However, this is not true. Acetylene is used widely in the manufacture of plastics and organic compounds, such as acetic acid and acetone. Most people who have experienced the odor of acetylene gas comment that it smells like garlic. This familiar odor comes from impurities, not from the acetylene. Pure acetylene, like natural gas, has no odor. The commercial acetylene used throughout the nation contains a small amount of impurities (less than one percent) consisting of air, phosphine, ammonia, and hydrogen sulfide.

In air, the flammable limits of acetylene range from 2.5 percent to 82 percent. This wide range of limits makes acetylene susceptible to ignition at almost any concentration.

When acetylene burns improperly in air, a considerable amount of black smoke is observed. However, when properly burned by combining with oxygen, acetylene burns with a bright flame and a temperature of about 5700°F.

The fact that acetylene has three bonds creates several problems, which means that this gas has to be treated differently from other compressed gases. One of the problems is instability. The three bonds together make acetylene unstable (highly reactive). If acetylene were compressed into standard cylinders like other gases are, it would detonate. Only 15 psig is the allowable, recommended safe limit for acetylene, instead of the normal 1500-2000 psig for most other compressed gases. Acetylene can be liquefied and solidified; however, it is even more dangerous in these forms. With increased concentration, it generally detonates because of its instability.

Storing Acetylene

The flammability and instability of acetylene prohibit its use in the compressed gas form. How, then, is it available in cylinders? To make this possible, manufacturers had to find a way not only to store the acetylene safely, but a way to store it in appreciable amounts.

The acetylene cylinder contains a "filler" to prevent pockets of acetylene from forming. If the cylinder is "shocked," the filler prevents the acetylene from detonating. In one type of construction, the cylinder is filled with calcium silicate. The calcium silicate filler is porous, and occupies only 8 percent of the cylinder. The pores are extremely small and, because of their size, inhibit flashback into the cylinder in the event of a fire. Filling the tank with acetylene alone does not solve the problem since at 15 psig only a small volume of acetylene could be used. To increase the volume, the acetylene is dissolved into an organic solvent, acetone. The acetone is used as an absorbent; one volume of acetone absorbs twenty-five times its volume of acety-lene. In this way, a large amount of acetylene can be stored in a safe and economical manner.

Safety devices for acetylene cylinders, which vary in capacities from 10 to 1,400 cubic feet, consist of fusible plugs. Depending on the size of the cylinder, the fusible plugs are found either on the top and/or bottom of the cylinder, with as many as four per cylinder. These fusible plugs melt at 212°F.

Acetylene will, under certain conditions, form explosive compounds with any available copper, mercury, or silver metals or salts. Therefore, contamination of acetylene with any of these should be avoided.

Hazards Connected With Acetylene Storage

If one is caught in the perilous situation of having to handle a leaking cylinder of acety-lene, the cylinder should be moved to open space, if possible. Caution must be taken in using a suitable open path in removal of the cylinder. Tipped cylinders, which are leaking gas and acetone should be righted, then removed.

Cylinders exposed to fire should be cooled. At no time should the area around the fusible plug be blocked or any part of the body be in line with it. The fire fighter must remember that 212°F is a low temperature when compared to ignited materials and must assume that the plug is going to melt.

Fires from acetylene tanks start in a variety of ways. Carelessness with cutting torches and placement of a cylinder too close to high heat areas have both been known to melt fusible plugs and ignite the acetylene. Controlled burning of the tank is advisable. With proper cooling of the tank (without extinguishment) the fire can be controlled until it burns itself out. Porosity size of the filler in the tank tends to inhibit the burning of acetylene back into the tank.

Fire Situations Involving Acetylene

In Santa Fe Springs, California, fire fighters had a potential catastrophe in progress on July 1, 1970, when they responded to a cryogenic plant where the acetylene filling area was involved. The fire started when a worker filled an acetylene tank. As he disconnected the tank, it erupted into flames, burning him. The fire, probably caused by static-electricity in combination with a leaking cylinder, spread rapidly.

Fire fighters were confronted not only with this danger on arrival, but had to protect exposures which consisted of liquid hydrogen tankers adjacent to the filling area, two storage tanks containing almost 90,000 gallons of liquid oxygen, an adjacent calcium carbide storage area, and a storage tank of propane. All this was done while the cylinders of acety-lene were burning, rupturing, and being hurled as far as 900 feet from the scene. No serious injuries, other than to the first worker, were reported. The fire fighters confined the blaze to the acetylene area only – averting a possible catastrophe – and kept the loss to $350,000.

In Winnipeg, Manitoba, fire fighters were faced with a similar situation in 1968 in a liquid-air manufacturing plant. The fire started after a worker emptied a drum of calcium carbide into an acetylene generator hopper. As he was preparing to load another drum, an explosion occurred and the fire started. Fire fighters responding were faced here with similar exposure problems – a calcium carbide storage area and a storage tank containing 2,000-3,000 gallons of liquid oxygen. The fire fighters managed to halt any spread of fire although exploding cylinders of acetylene were hurled through the air.

MAPP GAS

Increased concern over the instability of acetylene gas has brought interest in a substitute called MAPP gas: MAPP refers to methyl acetylene propadiene. MAPP is a stable liquefied petroleum gas. It is a mixture of methyl acetylene and propadiene and has a specific gravity of 1.48 compared to acetylene with 0.91.

MAPP gas has several advantages over acetylene. It is not shock sensitive, and thus can be stored in normal cylinders. Its explosive limits are like that of the normal flammable gases, not wide like acetylene. The limits range from 3.4 at the low to 10.8 for the high. Figure 14-1 compares these limits with those of acetylene and also with propane. MAPP gas is closer in properties to the other liquefied petroleum gases than it is to acetylene. It burns hotter in oxygen than the liquefied petroleum gases, as shown in table 14-5, but it burns some 400°F less than acetylene.

Although MAPP gas resembles the liquefied petroleum gases in most characteristics, it falls in between propane and acetylene in flame propagation. Fire fighting procedures are essentially the same as for propane and butane.

MAPP gas cylinders are filled to 85 percent of capacity with liquid and have a vapor pressure at 70°F of 94 psig. Pressure can be as high as 250 psig at 130°F. Acetylene, as previously discussed, must be kept at a pressure no greater than 15 psig. Another difference between MAPP and acetylene is the pressure-relief devices. Cylinders of MAPP gas are protected by a safety-relief valve which opens at 375 psig at approximately 175°F. Cooling the cylinder below 175°F resets the safety valve. The cylinders are also protected by frangible discs.

LIQUEFIED GASES

Some gases, because of their physical structure, become liquids when compressed. These gases are called the liquefied gases and

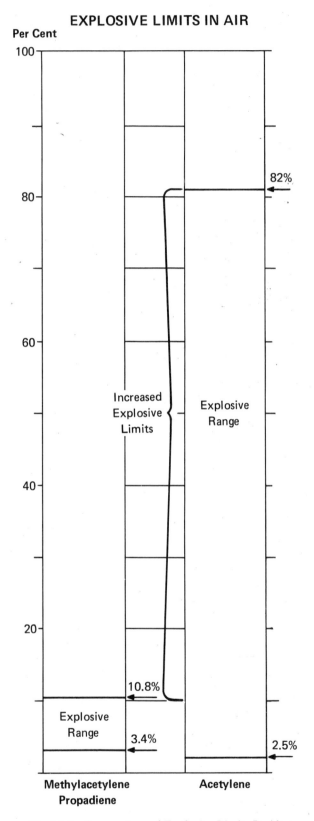

Fig. 14-1. Comparison of Explosive Limits In Air.

Gas	Temperature in °F
Natural Gas	4,600
Propane	4,600
MAPP	5,300
Acetylene	5,700

Table 14-5. Flame Temperature in Oxygen.

Gas	Critical Temperature °F	Critical Pressure psia
Ammonia	271	1657
Propane	206	617
Natural Gas	-116	673
Nitrogen	-232	493
Oxygen	-181	737
Hydrogen	-400	188

Table 14-6. Critical Temperatures and Pressures of Common Gases.

are quite common. The most commonly found liquefied gases are propane, butane, ammonia, chlorine, and refrigerant gases. By simply storing a gas in its liquid form, certain economical advantages are gained; one is volume.

Water, when heated above its boiling point, becomes a gas (steam). Steam, when brought below 212°F, becomes a liquid again in the form of water. Most other gases are the same. If they can be brought below their boiling point, they become liquids. Most people recognize this point. By storing a gas as a liquid, the important advantage of decreased volume is gained. For example, steam when it condenses decreases in volume some 1650 times, or one volume of water when it expands gives 1650 volumes of gas. When a gas is stored in the liquid form, a greater volume of the gas can be stored.

Gases can also be liquefied by applying pressure. However, there is a certain temperature at which a gas will not liquefy, no matter how much pressure is applied; this is called the *critical temperature*. For propane, the critical temperature is 206°F. It will not liquefy at any temperature above 206°F.

The pressure required to liquefy any gas at its critical temperature is called the critical pressure. For propane, this is 617 psig at a temperature of 206°F. As the temperature decreases, so does the pressure needed to

liquefy the gas. Table 14-6 lists the critical temperatures and pressures of the common gases.

The LP Gases

Two of the liquefied petroleum (LP) gases, propane and butane, are widely used for heating, drying, and recreational applications. Propane (C_3H_8) and butane (C_4H_{10}) are obtained from natural gas and from petroleum refinement.

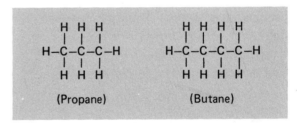

(Propane)   (Butane)

As shown, the two gases differ from methane in that they have three and four carbon atoms, respectively, compared to one for methane. This difference is significant since the vapor density of propane is 1.5 and of butane is 2.0, compared to 0.65 for methane. This means an additional hazard in the LP gases as compared to methane. They are heavier than air; and, therefore, tend to remain in lower levels. This means that they require different precautionary procedures than those

	Propane	Butane
Formula	C_3H_8	C_4H_{10}
Flash Point	Gas	Gas
Boiling Point	-44	31
Vapor Density	1.6	2.0
Critical Temperature (°F)	206	305
Critical Pressure (psia)	617	550
Flammable Limits % in Air		
Lower	2.1	1.8
Upper	9.5	8.4

Table 14-7. Properties of Propane and Butane.

used for methane. Table 14-7 shows additional properties of butane and propane.

Propane liquid boils at –44°F, while butane liquid boils at 31°F, close to the freezing point of water. This higher temperature limits the use of butane in northern areas of the United States where the temperature frequently goes below freezing. At this point, the butane liquid will not boil; and therefore, the gas flow stops.

To eliminate this problem, suppliers use propane predominantly or use mixtures of propane and butane. By using mixtures, the boiling point is dropped below 31°F to a temperature approaching –44°F, depending on the amount of propane used.

Hazards of LP Gases

Although nontoxic and noncorrosive, these two widely-used gases are flammable. Many incidents involving these two widely-used gases are recorded each year. Incidents involving propane have occurred at storage areas, on rail cars, and with tank trucks. They have involved leaks as well as fires.

For example, three tank cars, each with a 33,500 gallon capacity of propane, derailed after colliding with a switch engine in Ventura County, California, in November, 1973. The

cars, from a group of fourteen, had been released from a hold area by youths. They created a potential holocaust for fire fighters arriving at the scene.

Upon arrival, Ventura County fire fighters found a vapor cloud of propane spilling from the lead tank car. The derailed tank cars were in an embankment area and hid two injured rail employees. Fire fighters located and removed the men, who were partially frozen from the liquid propane, to an area where they could be evacuated to a hospital. Both eventually died.

Fire fighters took vapor readings while evacuation of the general area was in progress. The vapor readings indicated a high percentage of propane was present. Approximately 25,000 gallons of propane were estimated to have leaked from the damaged car. Remaining in the tank car were some 8,000 gallons. Water spray was used to heat the liquid propane and to cause increased vaporization. Later, when the spray became ineffective because of ambient temperature, heated water (120° F) was used until all the propane was dissipated. After this, the propane in the other two tank cars was transferred into additional tank cars; and the area was declared safe again.

Fire fighters in the town of Milford (East of Oneonta), New York, on February 12, 1974 were not as fortunate as Ventura County fire fighters, as a train derailment caused a tank car of liquid propane (approximately 30,000 gallons) to rupture and ignite. Seven tank cars derailed and eventually all were involved.

Arriving fire fighters attempted to cool the tanks near the burning tank car. However, because of the nature of the derailment, with cars piled on top of each other, efficient cooling was hampered. The safety-relief valves of some tanks began operating, causing ignited gas to heat adjacent tank cars. Eventually, the impinging flame and pressure buildup became too great, and a BLEVE[1] occurred. The BLEVE occurred immediately after the fire fighters were given the order to evacuate.

After the first BLEVE, there followed a second, third, and fourth. Parts of one tank car were hurled approximately one-quarter of a mile. Fifty fire fighters were injured, the majority from the first and second BLEVES. Two were burned critically.

Propane fires (and BLEVES) do not happen only in rail transportation. They can also happen in fixed storage locations. West St. Paul, Minnesota, fire fighters were faced with a BLEVE on January 11, 1974 in a storage tank (11,000 gallons) located between two apartment buildings.

The tank was being filled at 12:15 A.M. from a 10,000 gallon tanker when a spark occurred and fire engulfed the driver of the tanker. The driver was able to douse himself and call for help. Arriving fire fighters found a small fire under the tanker. Within minutes, this fire escalated into major proportions. The flame from the tanker impinged onto the vapor area of the storage tank. At the same time this was taking place, evacuation of surrounding apartment buildings was being completed.

The impinging jet of flame caused the storage tank to fail. Liquid propane spread over the area as the tank flew in many directions. One part of the tank hit a pumper, killing two fire fighters. The tank continued on and started one of the apartment houses on fire. More liquid propane started another apartment house on fire. Another portion of the tank flew and killed both a fire fighter and a woman fleeing from her apartment. In all, the fire killed four people, left two hundred twenty-five homeless, and did an estimated $3 million in damage.

Fighting LP gas fires was once thought to be mainly a concern for suburban and rural fire companies. These thoughts have changed because of an ever-increasing trend toward LP gas use in cities as a supplement to natural gas.

Handling Leaking LP Gas

Propane (and butane) leaks can be detected by odor. Although propane is odor free, mercaptan is added to give it a detectable odor. As liquid propane vaporizes, a vapor cloud is almost always present. This cloud is not propane. It is condensed moisture — condensed by the liquid propane at −44°F. Conditions for ignition depend upon the rate of gas discharge (flammable limits), wind velocity, and ambient temperature.

The problem for the person(s) handling the emergency is to stop the flow of leaking gas while eliminating possible ignition sources. Leaking vehicles or storage tanks present a formidable threat if ignition occurs. Dissipation of vapors from the immediate area is vital. No leaking propane tank should be approached without the back up of fog to dissipate the gas. In the event that a leak is caused by puncture, dispersion of the gas must be continued until the vapor readings indicate a safe condition.

Tanks in smaller sizes, such as those found around houses, should be righted if tipped. All leaks should be handled with extreme care since the temperature of the escaping liquid or gas is below zero. As a precaution, approach should always be upwind.

Controlling LP Fires

LP tanks are protected by relief valves which under fire conditions operate at approx-

---

[1] The word BLEVE means boiling liquid expanding vapor explosion. This condition occurs when heat builds up around the vapor area of the tank, from direct flame impingement on it, and weakens its structural strength. The boiling liquid vapors exert enough pressure on the weakened area to cause rupture. Depending on conditions, the tank may just split slightly with release of both liquid and vapor, or it may rupture violently with the tank being propelled a great distance.

200-400 psi, depending on the type and size of the tanks. The number of relief valves depends on the size of the tank.

Relief valves are a safety measure used to dissipate the increase in vapor pressure caused by the increase in heat. Tanks can rupture and do, when certain conditions occur. Cooling of tanks is of extreme importance: cooling results in a decrease of the pressure which eventually results in the containment of the fire.

Fires in large tank trucks or storage tanks which can be controlled by valve shutoff should be approached from both sides using 2 1/2-inch lines. The prime consideration is to cool the vapor area of the tank to prevent rupture from occurring. At the same time, a fog line should be used for radiant protection for the fire fighters.

After enough water is applied to the tanks for maximum cooling and the flame coming from the relief valve has subsided, the tank can be approached from the side, using two 2 1/2-inch lines for attack and one 2 1/2-inch line for backup (from another engine). The chief officer should direct the three lines, placing major emphasis on cooling the vapor area of the tanks and protecting the fire fighters.

The approach continues until a point is reached where opposing lines (if any are being used) must be shut down to avoid pushing the fire toward the advancing attack team. The attack team, as it gets closer, should switch from straight stream to fog stream for protection against the flame and radiant heat. Finally, the officer can close the valve after it is cooled with fog, and the fire fighters can back off slowly. The plan of attack of course differs with every department and depends heavily upon the local situation.

Preplanning is of extreme importance for fixed storage areas; without preplanning, a proper attack is difficult to carry out. Pre-planning is not possible, of course, in fires in tank vehicles where the leak has been caused through an accident and where no shutoff is possible. Such a fire must be controlled from master stream ranges, and with unmanned deluge sets, if possible. The tank(s) should be kept as cool as possible until all the fuel has been consumed.

CHLORINE GAS

Chlorine is a gas that deserves special attention. Although classified as a nonflammable gas with a green label, chlorine, because it is so widely used, has been involved in many health-related incidents.

Chlorine gas is greenish-yellow, and has a disagreeable, pungent odor. If inhaled, it is highly irritating and toxic. This gas is 2.5 times heavier than air and, as a result, lingers in low areas when present. This high specific gravity makes it difficult to dissipate the gas when leakage occurs and is the cause of most chlorine incidents.

Chlorine is found in areas where bacteria control is necessary, such as water and sewage treatment plants and swimming pools. Also, industrial plants dealing in any type of bleaching use large quantities of chlorine. Chlorine can also be found in manufacturing plants that make chlorine-containing compounds, such as trichloroethylene, pesticides, herbicides, and vinyl chloride. Chlorine is used as an intermediate in the synthesis of common "antifreeze" (ethylene glycol).

Fire Hazards of Chlorine Gas

Although classified as a nonflammable gas, chlorine can support combustion, and unites with certain organic and inorganic compounds with explosive violence. For example, a combination of hydrogen gas and chlorine gas explodes in the presence of any form of energy, such as light or heat. Chlorine even reacts with hot iron, causing it to burn.

The combination of chlorine with acetylene, ammonia, turpentine, and hydrocarbons can, and most probably will, cause an explosion. Not even the tanks themselves are safe from the action of chlorine gas. In combination with water and high heat (300°F or greater), chlorine attacks the cylinders. This is one of the reasons that water is not suggested for use on leaking chlorine; water and chlorine form hydrochloric acid, which is extremely corrosive.

Handling Chlorine Leaks

The following procedure is suggested for handling leaking chlorine cylinders. The fire fighters must wear self-contained breathing masks, preferably with positive pressure on the mask. If the gas cylinder is tipped, it must be placed upright to prevent the liquid from being expelled. If the cylinder has a hole in it, the tank must be positioned so gas and not liquid, escapes. (Sixty-six percent less chlorine will be formed.)

The chlorine gas distributor should be notified of the situation as soon as possible. If it is operable, the supply valve should be turned off. If not, caustic soda, soda ash, or hydrated lime should be used as a neutralizer. Caution: IF THE LEAK INVOLVES LARGE AMOUNTS OF CHLORINE, EVACUATE THE IMMEDIATE AREA.

Controlling Fires Near Chlorine

For fires near chlorine cylinders, the best possible procedure is to keep the cylinders cool. Water on a chlorine leak only makes it worse. Chlorine is only slightly soluble in water. The addition of water produces corrosive hydrochloric acid. Also, the water warms the tank allowing a greater flow rate of chlorine.

Shipping Chlorine

Chlorine is shipped in a variety of ways, mainly as a liquefied-compressed gas in cylinders, tank cars, and TMUs.

AMMONIA GAS

Ammonia is normally found in cylinders in the liquid-compressed gas state. Not only is it flammable, but it is also toxic. The eyes, skin, and mucous membranes are most generally affected when exposed to the vapor. In liquid form, ammonia causes severe burns on contact. Ammonia has a boiling point of –28°F at normal pressure. At this temperature, it is a colorless liquid.

Ammonia gas has a flammable range of 16-25 percent. Because of this higher level of flammability limits, the DOT lists it as green label. The gas, with a vapor density of 0.6, is almost one-half as light as air.

Ammonia is one of the gases used earliest for commercial refrigeration. It was used for years before halocarbons came into being. It is important not only for its use as a refrigerant, but also for its use in fertilizer synthesis. The ammonia can be added to the soil directly by injection, or it can be added through irrigation. It is used in the form of its salts, such as ammonium chloride, or in the form of ammonium nitrate. Ammonium nitrate itself does not burn, but detonates under extreme heat and pressure buildup. The Texas City disaster in 1947 testifies to this.

Ammonia Storage

There are two types of ammonia cylinders used for the liquid-vapor composition of ammonia. These are the bottle type and the tube type. Both are essentially of the same construction and differ only in the dip tube connected to them. The *dip tube* makes it possible for the user to withdraw either the liquid ammonia or the vapor. This is accomplished by placing the cylinder on its side and rolling it until the dip tube (which is about six to twelve inches in length) enters the liquid phase.

Shipping Methods

Ammonia is transported in cylinders, portable tanks, cargo tanks, and tank cars.

The tanks may be insulated or noninsulated. They are commonly equipped with spring-loaded safety-relief valves.

Handling Ammonia Leaks

The pungent odor of ammonia is easily detected. Exact location of a leak can be found in several ways. Litmus paper (red in color) is perhaps the easiest. This paper, when moist, turns from red to blue if it comes in contact with ammonia liquid or vapor. Hydrochloric acid may also be used. An open container of hydrochloric acid gives off white vapor when it is in the presence of ammonia gas. This white cloud, consisting of ammonium chloride, is easily visible.

REVIEW

1. A BLEVE is a

   a. Dike for holding liquids.
   b. Billowing low expanding vehicle explosion.
   c. Boiling liquid expanding vapor explosion.
   d. Tank on fire.

2. LP gases are in a

   a. Vapor-liquid relationship.
   b. Vapor-vapor relationship.
   c. Liquid-liquid relationship.
   d. Solid-liquid relationship.

3. Cylinders can be equipped with

   a. Frangible discs.
   b. Fusible plugs.
   c. Safety-relief valves.
   d. All of these.

4. All gases

   a. Burn.
   b. Support combustion.
   c. Expand indefinitely.
   d. Are unstable.

5. MAPP gas closely resembles which of the following in flammability limits?

   a. Acetylene.
   b. Propane.
   c. Oxygen.
   d. Chlorine.

6. Tube trailers transport gases

   a. In the liquefied form.
   b. In pressurized form.
   c. That are flammable only.
   d. None of these.

7. Color coding of cylinders

   a. Provides positive identification.
   b. Is an OSHA requirement.
   c. Is not necessary.
   d. Is a preliminary means of identification.

8. All liquefied gases have boiling points above

   a. 72°F.
   b. –150°F.
   c. 100°F.
   d. Ambient temperature.

9. Natural gas enters houses
   a. At high pressure.
   b. At moderate pressure.
   c. At low pressure.
   d. At no pressure.

10. Acetylene cylinders are protected by
   a. Safety-relief valves.
   b. Frangible discs.
   c. Fusible plugs.
   d. No safety device.

11. Class A poisons are protected by
   a. Safety-relief valves.
   b. Frangible discs.
   c. Soft plugs.
   d. No safety device.

12. Acetylene is not recommended to operate above
   a. 50 psig.
   b. 15 psig.
   c. 75 psig.
   d. 250 psig.

13. Chlorine is
   a. Toxic.
   b. A pungent gas.
   c. Corrosive.
   d. All of these.

14. Pressure required to liquefy any gas at its critical temperature is called
   a. Lower pressure.
   b. Critical pressure.
   c. Higher pressure.
   d. Ambient pressure.

15. Butane is not used in colder climates because of its
   a. Explosive limits.
   b. Availability.
   c. Cost.
   d. Boiling point.

16. Ammonia
   a. Is found only in the pressured form.
   b. Is found in a liquid-gas relationship.
   c. Has no flammable limits.
   d. Has no odor.

17. Oxygen is found
   a. In a pressurized state.
   b. In a liquid-vapor relationship.
   c. In hospitals.
   d. All of these.

18. Natural gas is
   a. Lighter than propane.
   b. Heavier than propane.
   c. Heavier than butane.
   d. Heavier than air.

19. Gases are characterized by their
   a. Low density.
   b. Ability to move freely.
   c. Lack of shape.
   d. All of these.

Unit 15 Cryogenic Gases

A Dutch physicist applied the name "cryogen" to certain gases that became liquids at extremely low temperatures. The term came from the Greek word "kryos," meaning icy cold. Icy cold indeed, for the science of cryogenic gases deals with gases turned into liquids at extremely low temperatures ($-150°F$ and below). These gases (hydrogen, oxygen, nitrogen, methane, and fluorine) along with the inert gases (helium, neon, argon, krypton, and xenon) have properties that are useful in various ways.

Electrical power can be distributed in a more economical fashion using cryogenic techniques. Food can be flash frozen to insure quality and good flavor. Steel mills and hospitals can operate more economically using liquid oxygen. The aerospace industry has gained sophistication through the use of liquid hydrogen, oxygen, and fluorine. Long term storage of blood has been made possible by flash freezing with liquid nitrogen. All these applications have brought increased use of the cryogenic liquids.

Today, cryogenic liquids are transported and stored throughout the United States. Gases that were formerly transported in a compressed form are now transported in a liquid form.

Gas	Boiling Point °F	Gas	Boiling Point °F
Helium	-452	Argon	-303
Hydrogen	-423	Oxygen	-297
Neon	-410	Methane	-259
Nitrogen	-320	Krypton	-244
Air	-315	Xenon	-163
Fluorine	-307		

Table 15-1. Boiling Points of Cryogenic Gases.

This means more gas per unit volume (200-700 times or more) is available. For example, a cubic foot of liquid oxygen expands to 842 cubic feet of gaseous oxygen when raised above its boiling point. Table 15-2 shows the expansion rates for a gallon of cryogenic liquid and also for a cubic foot of cryogenic liquid. All this adds up to increased economy in transporting and storage.

LIQUID OXYGEN

The uses of liquid oxygen are plentiful. It is found in research laboratories, medical laboratories, industrial plants, and in aerospace applications.

Liquid O_2, referred to as LOX, is obtained from the fractionation of air, and becomes liquid at $-297°F$. This temperature is slightly higher than liquid air, which has a boiling point of $-315°F$. The liquid oxygen is pale blue in color and is slightly heavier than water.

Hazards of Liquid Oxygen

Although classified as a nonflammable gas, liquid oxygen presents many hazards. Since it supports combustion, almost all combustible organic material burns with unbelievable intensity in its presence. Even many substances, generally considered noncombustible, burn ferociously in the presence of pure oxygen. This is why valves on oxygen cylinders are not oiled. Oil, being organic, would ignite readily in the presence of 100% oxygen. Even leaks of liquid oxygen onto asphalt (again organic) have caused combustion of the organic-based material.

Liquid oxygen is transported in trailers that resemble large thermos bottles. These trailers carry up to 10,000 gallons of LOX, and are clearly labeled LIQUID OXYGEN. They are protected with frangible discs and safety-relief valves. Storage tanks, vertical and cylindrical in structure, vary in capacity from 500 to 11,000 gallons of liquid.

Precautions

Liquid oxygen or any of the cryogenic gases are capable of "instant freezing" any part of the human body if contact is made with the liquid. Protective clothing should be worn to cover the body completely. Because of its porous nature, clothing retains gases such as oxygen. For this reason, any clothing that becomes contaminated should not be worn near any source of ignition. Clothing splashed with liquid should be removed. Since the liquids boil into large volumes of gas, a small amount of liquid inside clothing can cause a high area of oxygen-rich atmosphere, which makes the clothing easily ignitable.

Any problems involving liquid oxygen or any cryogenic material should be reported to the proper authorities for their advice in han-dling the situation. Flesh that comes in contact with the liquid should be washed with unheated water. The burn should be treated in a way similar to that for frostbite until qualified medical attention is available.

Fighting Fires Involving LOX

The method for fighting fires in oxygen-rich atmospheres is the same as any fire. However, a more intense and faster-burning fire than usual is encountered. To decrease the intensity of the fire, the flow of LOX should be stopped if possible.

Water is preferred as the extinguishing media for most fires. A spray is preferred to prevent the rapid boiling and splattering of liquid that may be caused by a straight stream. Water spray must never be directed on any part of the safety-relief devices. Freezing will occur and cause the safety-relief valves to be inoperative.

The following description of a LOX incident shows some of the problems associated with liquid oxygen. Three vehicles were destroyed as the result of a vapor cloud of

CUBIC FEET OF GAS PRODUCED

CRYOGENIC GAS	FORMULA	CUBIC FEET FROM 1 GALLON OF LIQUID	CUBIC FEET FROM 1 CUBIC FOOT OF LIQUID
Helium	He	101	737
Hydrogen	H_2	113	830
Neon	Ne	193	1414
Nitrogen	N_2	93	681
Air	—	97	712
Fluorine	F_2	128	936
Argon	Ar	112	823
Oxygen	O_2	115	842
Methane	CH_4	85	936
Krypton	Kr	93	546
Xenon	Xe	75	678

Table 15-2. Expansion Ratios – Liquid to Vapor.

pure oxygen formed from the release of some 600 gallons of LOX at the John F. Kennedy Space Center on March 25, 1970. Two vehicles were leaving a LOX storage area when they halted at the gate to await the arrival of a third vehicle. When the third vehicle arrived, the engines were shut off while a discussion was in progress. All three vehicles were in a "cloud" of vapor caused by the planned discharge of the LOX. When the drivers attempted to start their vehicles in the vapor area, smoke and flames started to come from the second car. At about the same time, the drivers of the first and third cars noted that their vehicles were also involved. Within minutes, the cars were completely engulfed and ruined![1]

Measurement of vapor concentration later indicated a concentration of 75-100 percent oxygen inside the vapor area. This high concentration of oxygen resulted in the increased burning rate of the vehicles.

The cause of the fire is unknown. However, speculation is that small amounts of gasoline vapor, in the presence of pure oxygen, were ignited by a possible spark from the starter motor.

LIQUID FLUORINE

Fluorine is the most reactive nonmetallic element. As a liquid at –306°F, it is slightly heavier than air with a 1.1 specific gravity. As a gas, it has a density of 1.7, as opposed to air at 1.0.

Fluorine is corrosive and extremely poisonous. It causes severe irritation burns to eyes and skin. As an oxidizer like oxygen, fluorine in combination with almost any substance, reacts vigorously and burns ferociously. Since fluorine even reacts with concrete and steel, fluorine containers are made of special steel with no impurities. Not only can cryo-

genic burns be obtained from fluorine but also chemical burns, which only complicate the healing process. Fluorine also reacts with water to form hydrogen fluoride, and oxygen.

$$2F_2 + 2H_2 \rightarrow 4HF + O_2$$

Fighting Fluorine Fires

Evacuate the area. Because of the health hazards involved, all fluorine leaks and fires should be handled with a minimum of manpower, when possible. DO NOT USE WATER directly on fluorine as it has been known to intensify the fire. Unmanned monitors and hose holders are recommended. Fire fighters must not have any parts of their bodies exposed when in the area of a fluorine leak.

LIQUID NITROGEN

Liquid nitrogen is colorless and odorless and is slightly lighter in weight than water. It is used mainly for refrigeration purposes. It is often found in laboratories where its inert properties are desirable.

Liquid nitrogen differs from liquid oxygen in that it is colder (–320°F) and that it does not support combustion. The fact that it does not support combustion reduces its hazard potential considerably. The temperature and its liquid-to-volume ratio are the principal dangers.

The fire fighter should be aware that liquid nitrogen (as all cryogenic liquids) freezes almost immediately anything with which it comes in contact. A banana placed in liquid nitrogen for one minute and then removed is hard enough to use to drive a nail into wood. This example clearly illustrates the potential danger associated with cryogenic gases.

A cubic foot of liquid nitrogen expands into 681 cubic feet of vapor when the liquid reaches its boiling point. If a sudden rupture of a tank occurs inside a closed area, the

---

[1] John B. Gayle, "Fire Incident in an Oxygen Cloud," *Fire Journal,* Jan. 1971, p. 78.

danger of asphyxiation is very real for the persons involved. The reason is the high liquid-to-vapor expansion ratio; a tank containing 6,000 gallons of liquid means a potential of 558,000 cubic feet of vapor.

LIQUID HELIUM

Liquid helium is the coldest of the cryogenic gases (–452°F). This temperature is only seven degrees from *absolute zero*, the point at which theoretically the molecules of a substance are at complete rest, or motionless. Liquid helium is used in medical applications as a cooling medium for nuclear reactors and as a protective layer for electronic work.

LIQUID NEON, ARGON, KRYPTON AND XENON

Liquid neon, argon, krypton and xenon are known as rare gases and are mainly used for laboratory experimentation. They are all colorless, odorless, and tasteless, with characteristics similar to liquid nitrogen. The most widely used of these gases is liquid neon, which is used mainly as a refrigerant and also for low temperature research.

LIQUID HYDROGEN

Hydrogen is the second coldest cryogenic gas with a boiling point of –423°F. Since it is flammable, extreme care must be exercised by anyone who is near the liquid. However, the hazard factor is somewhat reduced by the fact that liquid hydrogen dissipates readily, as it is very light.

A factor that should be mentioned is that open containers of cryogenic gases, such as helium, hydrogen, and nitrogen, are cold enough to freeze air and trap oxygen. In the case of hydrogen, the two elements (H_2 and O_2) present a formidable duo; explosion is likely to result. Care must be taken not to leave cryogenic containers open, causing condensation of the oxygen in the air.

LIQUID NATURAL GAS (LNG)

Natural gas, because of its wide usage, is in very great demand. The increasing use has prompted distributors to look to cryogenic storage tanks and cryogenic shipping vessels as a means of keeping up with the demand. In the past, huge storage tanks were filled with natural gas and acted as a means of storage. By going cryogenic, the same size storage tank can hold 623 times the amount of natural gas. This is not only economically sound but increases the supply to meet the demand. The greater demand also means an increasing number of fire departments will have LNG tanks situated in their districts as storage tanks in the near future.

There are two important factors to remember about natural gas in the liquid form. The boiling gas, as it forms vapor, is cold. At this stage, the gas is heavier than air and must be treated that way. Another factor is odor, which cannot be relied upon since the liquid is stored in the odor-free state. The odor-producing mercaptan is added after the liquid is vaporized and distributed. These factors are just the opposite of what the fire fighter is familiar with concerning vaporized natural gas.

Storage Facilities for LNG

LNG can be found in containers of various size and capacities. Cylindrical or spherical, these tanks have capacities from thirty-five million cubic feet to two billion cubic feet or greater. The smaller containers are designed for moderate pressures of 200 psi.

Leaking LNG

In a situation where LNG is leaking, control can best be established if the valve locations are known. Water can be used to dissipate the gas since water, at approximately

57°F, warms the vapors enough to disperse them. Care must be taken, as with any cryogenic liquid, that relief valves and pressure-protecting devices do not become iced up from water spray.

If LNG storage facilities or LNG transports are known to travel through a fire department jurisdiction, members of the department should learn about the behavior of LNG, both ignited and unignited. Representatives of the LNG industry have the information needed in most cases and should be contacted before an emergency arises.

Transportation of LNG

Liquefied natural gas is transported in the same manner as the other cryogenic gases. The tankers are composed of aluminum on the inside with a steel outer shell, separated by a vacuum between them (50-100 microns). The insulated tanks have a manual-relief valve and pressure-relief valves that open at 70 and 105 psi. The tanks are also equipped with frangible disks. The tanks have a capacity of some 11,600 gallons, which in cubic feet of gas, is approximately 80,000-90,000 cubic feet.

REVIEW

1. Cryogenic substances have a boiling point below minus
   - a. 100°F.
   - b. 150°F.
   - c. 200°F.
   - d. 250°F.

2. Liquid nitrogen is
   - a. Flammable.
   - b. Toxic.
   - c. Nonflammable.
   - d. None of these.

3. Liquid oxygen
   - a. Burns.
   - b. Is flammable.
   - c. Is toxic.
   - d. Supports combustion.

4. Fluorine is liquid at minus
   - a. 306°F.
   - b. 206°F.
   - c. 603°F.
   - d. 150°F.

5. Liquid fluorine reacts with
   - a. Metal.
   - b. Glass.
   - c. Water.
   - d. All of these.

6. Liquid hydrogen can condense
   - a. Tin.
   - b. Helium.
   - c. Boron.
   - d. Oxygen.

SECTION 4

COMBUSTIBLE MATERIALS

Unit 16 Metals

Metals, with the possible exception of mercury which is a liquid at ambient temperatures, are easy to recognize from their physical properties. By this time the student should have gained some insight also into the differences in the chemical properties of metals and nonmetals. Some characteristic properties of metals and nonmetals are summarized in table 16-1.

The best and probably the only practical way of understanding the hazardous properties of metals is by means of the periodic law correlations shown in figure 16-1, page 80. Of the total number of elements, eighty are metals and eighteen are nonmetals. Four elements having properties of both metals and nonmetals are called semimetals or metalloids.

The differences in the hazardous properties of the metals and nonmetals are due to the differences in the way they react with other elements. Nonmetals most readily fill their octet (outer shell of 8 electrons) by sharing electrons with other elements. Chlorine is an example of a typical nonmetal which forms chemical compounds by adding one electron to an outer layer of seven, thereby forming an octet. In contrast, a typical metal like sodium enters into chemical combination by transferring its only valence electron as shown in figure 16-2, page 80. Metallic character decreases and nonmetallic character increases with increasing numbers of electrons. Thus the elements with the highest metallic character are those to the left and top of the periodic table.

Physical Properties		Chemical Properties	
Metals	**Nonmetals**	**Metals**	**Nonmetals**
All are solids except mercury	May be solids, liquids, or gases	1-3 electrons in outer layer	4-8 electrons in outer layer
High density	Low density		
Metallic luster	No metallic luster	Have positive valence	Can have either positive or negative valence
Malleable	Nonmalleable		
Ductile	Brittle		
Good conductors of heat and electricity	Poor conductors of heat and electricity	React with water to form bases	React with water to form acids
Opaque	Transparent or translucent	Are reducing agents	Are oxidizing agents

Table 16-1. Comparison of the Properties of Metals and Nonmetals.

Group IA	II												III	IV	V	VI	VII	VIII
1 H		METALS					SEMIMETALS (METALLOIDS)				NONMETALS							2 He
3 Li	4 Be												5 B	6 C	7 N	8 O	9 F	10 Ne
11 Na	12 Mg												13 Al	14 Si	15 P	16 S	17 Cl	18 A
19 K	20 Ca	21 Sc	22 Ti	23 V	24 Cr	25 Mn	26 Fe	27 Co	28 Ni	29 Cu	30 Zn	31 Ga	32 Ge	33 As	34 Se	35 Br	36 K
37 Rb	38 Sr	39 Y	40 Zr	41 Nb	42 Mo	43 Tc	44 Ru	45 Rh	46 Pd	47 Ag	48 Cd	49 In	50 Sn	51 Sb	52 Te	53 I	54 Xe
55 Cs	56 Ba	57–71 La	72 Hf	73 Ta	74 W	75 Re	76 Os	77 Ir	78 Pt	79 Au	80 Hg	81 Tl	82 Pb	83 Bi	84 Po	85 At	86 Rn
87 Fr	88 Ra	89–103 Ac															

Fig. 16-1. Periodic Table Classification of Metals, Nonmetals, and Semimetals.

THE ALKALI METALS

The most reactive metals are the alkali metals which occur in Group I of the periodic table. Selected physical and chemical properties are shown in table 16-2. All are intensely reactive with oxygen and water. The hazards associated with the formation of caustic hydroxides and caustic oxides, and with hydrogen liberation were mentioned earlier. The alkali metals have low melting points which also adds to their hazard.

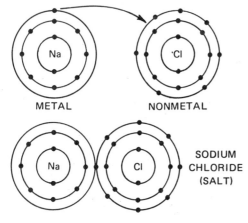

Fig. 16-2. Reaction of a Metal With a Nonmetal.

Areas of Use

Lithium, sodium, and potassium are commonplace reagents in most laboratories (ranging from high school to industrial level). Larger quantities may be encountered in industrial and transportation accidents. They are stored under kerosene to exclude moisture and oxygen. Most accidents occur because the metals are removed from this protective environment with the subsequent ignition of the hydrogen formed by reaction with water.

Metallic sodium is used in large quantities where a strong reducing agent is required as in the manufacture of organic dyes, drugs, and in the reduction of salts of rarer metals to the free metals. It is also widely used as a dehydrating agent in the preparation of anhydrous (dry) solvents. For example, anhydrous ether can be obtained by adding metallic sodium. This drying process depends on the reactivity of sodium with water.

The largest industrial users of metallic sodium are the manufacturers of sodium cyanide (NaCN), sodium peroxide (Na_2O_2) and

Metal	Symbol	MP, °C	BP, °C	Reactive With			Distribution
				O_2	H_2O	Halogens	
Lithium	Li	180	1326	Yes	Yes	Yes	Common in laboratories
Sodium	Na	98	889	Yes	Yes	Yes	Common in laboratories and industry
Potassium	K	63	757	Yes	Yes	Yes	Common in laboratories
Rubidium	Rb	39	679	Yes	Yes	Yes	Rare
Cesium	Cs	29	690	Yes	Yes	Yes	Rare

Table 16-2. Physical and Chemical Properties of the Alkali Metals.

tetraethyl lead $[Pb(C_2H_5)_4]$. Sodium is extensively used in the manufacture of metal alloys to remove unwanted impurities such as oxygen, sulfur, and halogens. Other industrial applications of sodium are in electrical conductors where large currents must be carried for short distances, in heat conductors as in aircraft engine exhaust valves and nuclear reactors, and in sodium vapor lamps.

The industrial applications of sodium indicate industries in which sodium can be expected to be encountered. Since sodium is the cheapest of the alkali metals it is the most widely used. For the same reason, lithium and potassium are most likely to be encountered on a smaller scale in laboratories. The possible exception is the limited use of lithium as a source of a brilliant red color in pyrotechnics.

The other alkali metals, rubidium and cesium are quite rare and unlikely to be encountered in quantities greater than a few pounds. The reactivity properties are similar to sodium except that they are more *pyrophoric* (ignite spontaneously) with air. They do not have any distinctive properties which make them particularly suitable in specialty applications.

The alkali metals are the most reactive of the metals. Oxygen, on the other hand, is not the most reactive of the nonmetals. The *halogens* F, Cl, Br, and I in Group VII of the periodic table are much more reactive. It follows that the hazards associated with reaction of alkali metals with halogens are greater than with oxygen. The rate of reaction is faster, more heat is evolved, and the salts of fluorine and bromine are toxic. The alkali metal portion of salts (Na in NaF, for example) is not toxic.

Fire Fighting Procedures

Extreme care should be taken when the alkali metals are burning or exposed. Never use water in the presence of these metals since hydrogen is given off along with a large amount of heat.

$$2Na + 2HOH \rightarrow 2NaOH + H_2 + heat$$

The heat is sufficient to cause the hydrogen to ignite, causing an explosion with the splattering of caustic. Class D extinguishers (Met-L-X, TEC, Pyrene G-I) containing graphite and sodium chloride are recommended for use. In cases of extreme emergency, sand can be used — but only if it is dry. Moist sand can cause steam explosions.

- Foam, carbon dioxide or halogenated extinguishers should never be used on metal fires. They produce a violent reaction.

- Alkali metals must never be handled with bare hands. The metals are so reactive that they react with the moisture in the hands, causing severe burns. Full protective equipment must be worn when dealing with these metals.

Metal	Symbol	MP, °C	BP, °C	Reactive With			Distribution
				O_2	H_2O	Halogens	
Beryllium	Be	1280	3000	Yes	Yes	Yes	Common in industry
Magnesium	Mg	650	1100	Yes	Yes	Yes	Common in industry
Calcium	Ca	850	1490	Yes	Yes	Yes	Common in industry
Strontium	Sr	770	1384	Yes	Yes	Yes	Pyrotechnics industry
Barium	Ba	710	1638	Yes	Yes	Yes	Electron tube industry
Radium	Ra	700	1140	Yes	Yes	Yes	Rare

Table 16-3. Physical and Chemical Properties of the Alkaline Earth Elements.

- Since these metals react with the moisture in the air, they are shipped under kerosene or mineral oil. If the liquid is involved in a flammable liquid fire use CO_2 to extinguish. **DO NOT USE** water. Water will sink to the bottom, react with the metal, and cause violent reactions — making conditions worse.

ALKALINE EARTH METALS

The next clearly defined family of metals are the alkaline earth metals in Group II of the periodic table, table 16-3. Just as an alkali metal loses one electron to form a singly charged ion, the alkaline earth metals lose two electrons to form a doubly charged ion.

alkali metal	$Na \rightarrow Na^+ + e$
alkaline earth metal	$Ca \rightarrow Ca^{++} + 2e$

Since the nuclear charge is one greater (+2 as compared to +1), the two valence electrons in the alkaline earth elements are held more strongly than the single valence electron of the alkali metals. The alkaline earth metals release electrons less readily than the alkali metals and for that reason are less reactive. They react with oxygen, water, and halogens as sodium does, except that the reaction is less vigorous. The alkaline earth elements are widely used and the probability of encountering them in hazardous situations is reasonably high.

Areas of Use

Beryllium is used in beryllium alloys and in beryllium oxide ceramics (BeO). Magnesium is used in aircraft and marine construction, in photographic flashbulbs, and in incendiary bombs, flares, and pyrotechnics; it is found in the engine blocks of some imported vehicles. Calcium is used as a deoxidizer in metallurgy and in alloys such as bearing metals. Strontium provides the crimson color in pyrotechnic devices. Barium is used as a getter in the manufacture of electron tubes. Radium is used in cancer treatment.

Fire Fighting Procedures

Some of the alkaline earth metals are highly toxic, particularly beryllium and barium. Both are primarily heavy metal poisons.

Metal	Symbol	MP, °C	BP, °C	Reactive With			Distribution
				O_2	H_2O	Halogens	
Aluminum	Al	660	2057	Low	Low	Yes	Common
Gallium	Ga	30	1983	Low	Low	Yes	Rare
Indium	In	156	2000	Low	Low	Yes	Rare
Thallium	Tl	302	1460	Low	Low	Yes	Rare

Table 16-4. Physical and Chemical Properties of the Aluminum Family of Metals.

Classification	Metal	Symbol	MP, °C	BP, °C	Reactivity	Toxicity	Distribution
Group IV	Germanium	Ge	959	2700	low	- - -	electronic semiconductors
	Tin	Sn	232	2270	low	- - -	cans, electroplating
	Lead	Pb	327	1620	low	high	solder, paint
Group V	Antimony	Sb	631	1440	high	high	lead storage batteries
	Bismuth	Bi	271	1420	high	high	low melting alloys
Group VI	Polonium	Po	- - - -	- - - -	high	high	radioactive, rare
Transition Metals							
Titanium Family	Titanium	Ti	1800	5100	high temp.	- - -	paint, low density metal
	Zirconium	Zr	1857	5050	high temp.	- - -	ceramics
	Hafnium	Hf	2500	3700	high temp.	- - -	rare
Vanadium Family	Vanadium	V	1710	- - - -	high temp.	- - -	alloying element
	Niobium	Ni	2500	3300	high temp.	- - -	alloying element
	Tantalum	Ta	2996	6100	high temp.	- - -	chemical, electronic industry
Chromium Family	Chromium	Cr	1890	2480	high temp.	- - -	alloying element
	Molybdenum	Mo	2620	4800	high temp.	- - -	alloying element
	Tungsten	W	3370	5900	high temp.	- - -	alloying element, light filaments
Manganese Family	Manganese	Mn	1220	2150	high temp.	high	alloy steels
	Technetium	Tc	- - - -	- - - -	high temp.	- - -	rare
	Rhenium	Re	3170	- - - -	high temp.	- - -	rare
Iron Family	Iron	Fe	1535	3000	moderate	- - -	widespread
	Ruthenium	Ru	2450	2700	high temp.	- - -	rare
	Osmium	Os	2700	5300	burns in air	high	alloys
	Cobalt	Co	1490	2890	low	- - -	alloys
	Rhodium	Rh	1970	2500	low	- - -	catalyst, electroplating
	Iridium	Ir	2454	4800	low	- - -	catalyst
	Nickel	Ni	1452	2732	low	- - -	catalyst, alloys
	Palladium	Pd	1549	2200	nil	- - -	catalyst
	Platinum	Pt	1773	4300	nil	- - -	catalyst
Copper Family	Copper	Cu	1083	2595	moderate	- - -	widespread
	Silver	Ag	961	2212	moderate	- - -	widespread
	Gold	Au	1063	2966	low	- - -	jewelry
Zinc Family	Zinc	Zn	419	907	high	- - -	galvanized iron
	Cadmium	Cd	321	767	high	high	electroplating
	Mercury	Hg	-39	357	low	high	instruments
Lanthanide Family					high	- - -	rare
Actinide Family					radioactive	high	rare

Table 16-5. Summary of the Properties of the Transition Metals and Remaining Groups IV, V, and VI Metals.

Beryllium is also especially hazardous as a respiratory poison and causes a lung disorder known as beryllosis. Radium is a hazardous material primarily because of its radioactivity.

The same procedures apply to these metals as to the alkali metals. The Group II metals

Name	Symbol	MP, °C	BP, °C	Reactivity With			Distribution
				O2	H2O	Halogens	
Boron	B	2300	2550 (sublimes)	Yes	Yes	Yes	Rare as metal
Silicon	Si	1420	2355	Yes	Yes	Yes	Electronic semiconductors
Arsenic	As	814	615	Yes	Yes	Yes	Rare as metal
Tellurium	Te	452	1390	Yes	Yes	Yes	Rare as metal

Table 16-6. Physical and Chemical Properties of the Metalloids.

are not as reactive as Group I. Water on magnesium only hastens its burning, a fact known to anyone who has experienced magnesium engines burning on some foreign cars. Class D extinguishers (Met-L-X, TEC, Pyrene G-I) are recommended for extinguishment. Handling metals such as calcium is not advisable unless protective gloves are used. Calcium causes skin burns the same as sodium does.

METALLIC ELEMENTS

The metallic elements in Group III of the periodic table are summarized in table 16-4, page 82. Boron has the transition properties of a semimetal and will be discussed with the other semimetals later.

The elements of the aluminum family are less reactive than the alkaline earth elements. This follows because the nuclear charge is +3 and the three valence electrons are held more strongly than the two electrons of the alkaline earth metals.

Area of Use

Of the five elements in this family, only boron and aluminum are of much industrial significance. Boron, which is largely nonmetallic in its properties, is important primarily for its compounds. Aluminum is a very widely used metal which hardly needs elaboration. The other elements of this family have only limited commercial applications. Gallium is used in high temperature thermometers. Indium is used in electroplating bearing metals

and jewelry. Thallium has no use as a metal. Its only application is in rat and ant poisons as the sulfate salt $[Tl_2(SO_4)]$.

Fire Fighting Procedures

The Group III metals are active only when in a dust or powdered form. Procedures to be followed are the same as for fighting any class D fire.

Handling of these metals does not cause burns as the Group I and II metals do. Again, water, foam, dry chemical, halogenated extinguishers and carbon dioxide are **not** to be used.

Of the remaining metals, none have a significantly high flammability hazard although many of them are highly toxic. The hazardous aspects and industrial applications of the remaining metals are summarized in table 16-5 and those of the semimetals in table 16-6.

Many of the low melting alloys used in industry contain high concentrations of toxic metals. Since their composition cannot be seen in the trade name, some of the more common toxic alloys are summarized in table 16-7.

Name	Composition, %
Dentist amalgam	Hg 70, Cu 30
Solder, lead	Pb 67, Sn 33
Solder, silver	Ag 63, Cu 30, Zn 7.0
Battery plate	Pb 94, Sb 6
Wood's metal	Bi 50, Cd 12.5, Pb 25, Sn 12.5
Type metal	Pb 70, Sb 18, Cu 2

Table 16-7. Composition of Some Common Alloys Containing Toxic Metals.

FACTORS AFFECTING REACTIVITY OF METALS

Many of the metals are inert at ambient temperatures but react at higher temperatures with nonmetals (including carbon, hydrogen, oxygen, HCl and SO_2). Reaction at high temperatures can produce metal oxides or other salts which are highly toxic.

Usually, hazards associated with the combustion of relatively inert metals do not occur with massive pieces of metal. Finely divided powders of any metal, however, can be rendered explosively combustible if their particle size is small enough. This is true of plastics, wood, coal, flour, or any other incompletely oxidized material. In contrast, finely powdered sand, SiO_2, is not combustible in air because it is fully oxidized. In special environments, such as hydrogen or fluorine, finely divided sand does react explosively.

The reactivity of metal powders is a primary reason for the rather extensive discussion of metals which are inert under ordinary conditions. Platinum and other "inert" metal powders are used extensively in the petroleum refining industry in catalytic cracking and gasoline platforming processes. Although it seems intuitively wrong to picture platinum powder as a combustion hazard, these metal powder catalysts are often highly combustible and also present a toxicity hazard.

REVIEW

1. Metallic character of an element increases

   a. As the number of valence electrons increases.
   b. As the number of valence electrons decreases.
   c. As the solid metal forms from the liquid metal.
   d. In going from Group I to Group IV of the periodic table.

2. A typical example of a metal is

   a. Barium.
   b. Silicon.
   c. Radon.
   d. Arsenic.

3. All metals are

   a. Solid.
   b. Oxidizing agents.
   c. Reducing agents.
   d. Toxic.

4. When a metal reacts

   a. It transfers an electron to a nonmetal.
   b. It gains an electron from a nonmetal.
   c. It forms a compound with a lower melting point.
   d. It is no longer hazardous.

5. The most reactive metals are

   a. Aluminum and magnesium.
   b. Sodium and calcium.
   c. The alkali metals.
   d. The alkaline earth metals.

6. The alkali metal with the highest metallic character is

   a. Lithium.
   b. Sodium.
   c. Magnesium.
   d. Calcium.

7. The alkali metal most likely to be encountered is sodium because

   a. It is the strongest reducing agent.
   b. It is less toxic than lithium.
   c. It is the least hazardous.
   d. It is the least expensive.

8. The hazards associated with the reaction of sodium are greatest with

   a. Oxygen.   c. Chlorine.
   b. Fluorine.   d. Hydrocarbons.

9. The alkaline earth metals are less hazardous than the alkali metals because

   a. They are less toxic.
   b. They are not encountered as frequently.
   c. They release electrons less readily.
   d. They have fewer valence electrons.

10. The most toxic alkaline earth metal is

    a. Magnesium.   c. Strontium.
    b. Beryllium.   d. Calcium.

11. Strontium reacts with water to form

    a. An oxide.   c. An acid.
    b. A base.   d. A toxic compound.

12. The toxic heavy metal, lead, is most likely to be encountered in

    a. Medical laboratories.   c. Low melting alloys.
    b. Dental offices.   d. Thermometers.

13. The toxic heavy metal, mercury, is most likely to be encountered in

    a. Pyrotechnics.   c. Dentist amalgam.
    b. Batteries.   d. Electroplating shops.

14. Which of the following metals are toxic?

    a. Antimony, osmium, and manganese.
    b. Germanium, barium, and radium.
    c. Potassium, lithium, and calcium.
    d. Titanium, lead, and mercury.

15. Antimony is most likely to be encountered in

    a. The petroleum industry.   c. The battery industry.
    b. The aircraft industry.   d. The pharmaceutical industry.

16. Osmium is most likely to be encountered in an

    a. Electroplating shop.   c. Electronics industry.
    b. Alloy industry.   d. Solders.

17. Which of the following are most apt to be hazardous?

    a. Low melting alloys.   c. Transition metals.

    b. High melting alloys.   d. Metalloids.

18. Platinum powder is most likely to be encountered in

    a. Jewelry stores.   c. Oil refineries.

    b. Banks.   d. Pyrotechnic plants.

19. Metal powders are hazardous because

    a. They have more surface area than the solid metal.

    b. They are more exothermic than the solid metal.

    c. They are stronger reducing agents than the solid metal.

    d. They are more endothermic than the solid metal.

Unit 17 Nonmetals

The chemistry fundamentals necessary for discussion of the hazards associated with nonmetals were established in the previous unit and need not be repeated here. The task is also considerably simpler because of the greater familiarity with the behavior of nonmetals and the fact that the number to be dealt with is much smaller. The nonmetals are

Group IV	Carbon
Group V	Nitrogen and Phosphorus
Group VI	Oxygen, Sulfur, and Selenium
Group VII	The Halogens: Fluorine, Chlorine, Bromine, and Iodine
Group VIII	The Inert Gases: Helium, Neon, Argon, Krypton, and Xenon

CARBON

Carbon is not ordinarily thought of as a hazardous material. However, if it were not for carbon and its compounds, there would be little need for fire departments. In this unit, only elemental carbon is considered. The compounds of carbon are discussed later. The various forms of carbon which are commonly encountered are summarized in table 17-1. All are hazardous because they are very porous and have a high surface area. The effect is analagous to that created by finely divided metals. Combustion can proceed very rapidly or an explosion can result. Also, the reaction of carbon with oxygen is very exothermic and very high flame temperatures are achieved.

Carbon can be oxidized by elements other than oxygen. All of the reactions are highly exothermic and the resultant products are hazardous.

By sulfur:	$C + 2S \rightarrow$	CS_2 carbon disulfide
By fluorine:	$C + 2F_2 \rightarrow$	CF_4 carbon tetrafluoride
By chlorine:	$C + 2Cl_2 \rightarrow$	CCl_4 carbon tetrachloride
By bromine:	$C + 2Br_2 \rightarrow$	CBr_4 carbon tetrabromide

PHOSPHORUS

Phosphorus occurs in several allotropic forms as shown in table 17-2. Different forms of the same element are known as *allotropes.* Allotropes of the elements occur because the octet of electrons which determines chemical behavior can sometimes be arranged to attain reasonable levels of stability in more than one way. Allotropes are very common among the nonmetals: notably carbon, oxygen, sulfur and phosphorus.

Oxygen (O_2) and ozone (O_3) are allotropes.
Carbon (C) and graphite (C_6) are allotropes.

White phosphorus is obtained when phosphorus vapors are condensed to the solid.

Type of Carbon	Source and Properties
Charcoal	Heating of wood or bones in absence of air to form porous carbon structure.
Coal	Slow decay of wood under pressure in earth. The result is similar to charcoal. The higher pressures in the interior of the earth yield a less porous structure.
Carbon black	Soot formed in various combustion processes in which the amount of oxygen is insufficient to promote complete combustion.
Graphite	Carbon arranged in a specific six-sided crystal structure.

Table 17-1. Forms of Carbon.

Form	MP, °C	BP, °C	Solubility	Hazard
White	44	280	soluble in carbon disulfide, benzene, ether	high
Red	600	- - -	insoluble	moderate
Violet	620	- - -	insoluble	moderate
Black	- - -	- - -	insoluble	moderate

Table 17-2. Allotropes of Phosphorus.

In this form it melts at 44°C. It looks like a soft, waxlike material, usually in the form of sticks, with a yellowish coating. The white form is so reactive with oxygen that safety demands that it is stored under an inert solvent. As was pointed out in unit 12, phosphorus does not react with water at ambient temperatures. Therefore, water can be used as the "inert" solvent. Phosphorus does react with water at elevated temperatures, however.

When white phosphorus is heated in the absence of oxygen to about 245°C, it changes to *red phosphorus.* Small crystals form, the density is increased and the highly hazardous white form is changed to the relatively safe red form. Red phosphorus is not pyrophoric with air as is the white form. Red phosphorus is not stored under an inert solvent. *Violet phosphorus* is believed to be chemically the same as red phosphorus but differs in that the crystal size is larger.

Black phosphorus is formed when white phosphorus is subjected to pressure. A black allotrope is formed which is somewhat less reactive than the red form, presumably because of its higher density.

Area of Use

White phosphorus is a widely used basic material and can be encountered in a number of situations. It is used as a basic material in laboratories and industry, as an alloying element (as in phosphorus bronzes), for laying down smoke screens because of the dense white cloud of oxide formed when phosphorus is burned in air. This, incidentally, is a highly hazardous type of smoke screen because the oxide (P_2O_5) reacts with water in the lungs to produce phosphoric acid (H_3PO_4). Matches (the nonsafety variety) are made from phosphorus sesquisulfide (P_4S_3).

Phosphorus reacts with most of the same types of materials as carbon does.

With oxygen: $4P + 5O_3 \rightarrow 2P_2O_5$ phosphorus pentoxide

With sulfur: $4P + 3S \rightarrow P_4S_3$ phosphorus trisulfide

With fluorine: $2P + 5F_2 \rightarrow 2PF_5$ phosphorus pentafluoride

With chlorine: $2P + 5Cl_2 \rightarrow 2PCl_5$ phosphorus pentachloride

With bromine: $2P + 5Br_2 \rightarrow 2PBr_5$ phosphorus pentabromide

Fire Fighting Procedure for Phosphorus

Fire fighting in situations involving phosphorus centers around excluding oxygen. Water is effective because it cools and functions as an "inert" material. The problem is that phosphorus ignites again when the water evaporates, as it eventually will. For this reason, the phosphorus fire must be extinguished, and the phosphorus must then be moved to a location where it can be burned safely.

SULFUR

Sulfur, like phosphorus, has several allotropic forms, but none are important with regard to hazardous properties. Elemental sulfur is inert to oxygen and water at ambient conditions. There is no difference among the allotropic modifications.

Sulfur combines with all of the elements except the platinum metals, gold, and the inert gases — much as oxygen does.

$$2H_2 + O_2 \rightarrow 2H_2O$$
hydrogen oxide

$$C + O_2 \rightarrow CO_2$$
carbon dioxide

$$4Na + O_2 \rightarrow 2Na_2O$$
sodium oxide

$$H_2 + S \rightarrow H_2S$$
hydrogen sulfide

$$C + 2S \rightarrow CS_2$$
carbon disulfide

$$2Na + S \rightarrow Na_2S$$
sodium sulfide

The reactions of sulfur with metals such as copper and iron are so vigorous that the heat evolution raises the temperature to incandescence. (Selenium is a highly toxic nonmetal which reacts in the same way as oxygen and sulfur, but because of its relative rarity will not be considered in detail here.)

Hazards of Sulfur

The hazards of sulfur and sulfur compounds are primarily associated with the formation of toxic gases. In the usual oxidizing atmosphere, sulfur dioxide is formed. Conversely, in reducing atmospheres hydrogen sulfide is formed. Both sulfur dioxide and

$$S + O_2 \rightarrow SO_2 \quad \text{sulfur dioxide}$$
$$S + H_2 \rightarrow H_2S \quad \text{hydrogen sulfide}$$

hydrogen sulfide are highly toxic gases which can cause death to persons exposed to parts per million concentrations.

THE HALOGENS

The remaining nonmetals, the halogens, are all highly hazardous oxidizing agents. The most reactive of the halogens, fluorine and chlorine, react directly with all of the metals to form metal halides. Bromine and iodine are somewhat less reactive but nevertheless form metal halides with all the metals with less violent consequences. The comparative reactivity of oxygen in most chemical reactions lies between that of iodine and bromine.

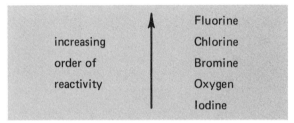

increasing	Fluorine
order of	Chlorine
reactivity	Bromine
	Oxygen
	Iodine

(increasing order of reactivity ↑)

The halogens react with water in two ways. In the following reactions, the halogens are represented by the symbol X. This is called a displacement reaction because the oxygen of H_2O is replaced by a halogen.

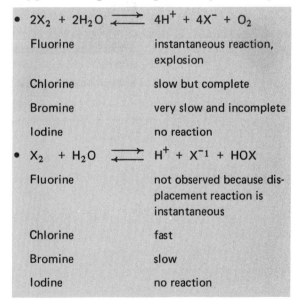

- $$2X_2 + 2H_2O \rightleftharpoons 4H^+ + 4X^- + O_2$$

Fluorine	instantaneous reaction, explosion
Chlorine	slow but complete
Bromine	very slow and incomplete
Iodine	no reaction

- $$X_2 + H_2O \rightleftharpoons H^+ + X^{-1} + HOX$$

Fluorine	not observed because displacement reaction is instantaneous
Chlorine	fast
Bromine	slow
Iodine	no reaction

Hazards of the Halogens

Fluorine is highly hazardous because of its high reactivity. Both chlorine and bromine are stronger oxidizing agents than oxygen and are relatively more hazardous.

	Fluorine	Chlorine	Bromine	Iodine
Appearance	Pale yellow gas/ liquid, pungent odor	Greenish-yellow gas, pungent odor	Dark reddish-brown liquid	Greyish-black solid; vapors are violet
Boiling Point: ($^\circ$F)	–306	–30	137	365
Density gas/liquid:	1.695	1.56	3.5	- - - -
Reactivity:	High	Moderate	Moderate	Low
Hazard:	Highly toxic, corrosive	Highly toxic	Highly toxic	Highly toxic
Shipping Regulations:	Red gas label	Green label	Corrosive Liquid white label	Poison label

Table 18-3. The Halogens.

- All of the halogens are intensely irritating to the mucous membranes and respiratory tract. Initial symptoms are those of pneumonia. Prolonged inhalation can have serious consequences. All are strong oxidizing agents and can cause severe flesh burns.

- Approach all halogens with extreme care. Fluorine should not be approached; the area should be evacuated until properly trained and clothed personnel arrive at the scene. Fluorine and chlorine were discussed more thoroughly in units 14 and 15. Table 17-3 lists the properties of the halogens.

REVIEW

1. The nonmetallic character of an element increases
    a. When a metal reacts with oxygen to form an oxide.
    b. In going from Group VII to Group IV of the periodic table.
    c. As the number of valence electrons decreases.
    d. As the number of valence electrons increases.

2. Which of the elements is a nonmetal?
    a. Selenium.
    b. Silicon.
    c. Arsenic.
    d. Germanium.

3. An example of a nonmetal is
    a. Water.
    b. Sodium vapor.
    c. Oxygen.
    d. All of these.

4. White phosphorus is
    a. Hazardous.
    b. Exothermic.
    c. An allotrope.
    d. All of these.

5. All allotropes are
    a. Hazardous.
    b. Different forms of the same element.
    c. Same forms of different elements.
    d. Different forms of different elements.

6. Phosphorus smoke screens are hazardous because
    a. Visibility is limited.
    b. They are used in military applications.
    c. The smoke is toxic.
    d. They are exothermic.

7. In fires involving phosphorus, the most important preventative measure is to
    a. Avoid contact with water.   c. Avoid contact with carbon.
    b. Avoid contact with alkali metals. d. None of these

8. White phosphorus is most commonly stored under
    a. Water.   c. Carbon dioxide.
    b. Kerosene.   d. Helium.

9. Sodium is most commonly stored under
    a. Water.   c. Carbon dioxide.
    b. Kerosene   d. Helium.

10. Nonmetals react with
    a. Metals.   c. Both metals and nonmetals.
    b. Other nonmetals.   d. Semimetals only.

11. The chemical reactions of sulfur are most similar to those of
    a. Phosphorus.   c. Hydrogen.
    b. Carbon.   d. Oxygen.

12. In a reducing atmosphere, sulfur forms the toxic combustion product
    a. SO_2.   c. H_2S.
    b. SO_3.   d. H_2SO_4.

13. The hazardous chemical reactions of fluorine are most similar to those of
    a. Oxygen.   c. Sulfur dioxide.
    b. Chlorine.   d. Bromine.

14. Fluorine reacts with water to produce
    a. Heat.   c. Acidic products.
    b. Hydrogen fluoride.   d. All of these.

15. Which of the following is the strongest oxidizing agent?
    a. Oxygen.   c. Iodine.
    b. Bromine.   d. Sulfur dioxide.

16. In most chemical reactions, chlorine is more reactive than
    a. Iodine.   c. Bromine.
    b. Oxygen.   d. All of these.

17. All the halogens have distinctive
    a. Colors.   c. Vapor densities.
    b. Odors.   d. Flash points.

Unit 18 Hydrocarbons

Hydrocarbons, as the name implies, are compounds of hydrogen and carbon. Carbon atoms have an amazing tendency to share electron pairs with one another to form an almost infinite variety of compounds. Thus, carbon with four electrons in the outer orbit can form linear, branched, and ring compounds by simply sharing pairs of electrons.

linear	branched	ring
C:C:C:C	C •• C •• C:C:C:C:C •• C •• C	C C C C C C

When hydrogen atoms are added to complete the octet of electrons, the result is a hydrocarbon.

linear

H H H H
H:C:C:C:C:H
H H H H

branched

H
••
H:C:H
••
H:C:H
••
H:H •• H:H
H:C:C:C:C:C:C:H
H H •• H H
H:C:H
••
H:C:H
••
H

ring

H   H
C
H:C:H   H:C:H
H:C:H   H:C:H
C
H   H

All hydrocarbons contain carbon atoms with an oxidation number of four and hydrogen with an oxidation number of one. In that respect, hydrocarbons are simple. The complexity results from the large number of different compounds which can be formed and the associated problem of naming the many compounds.

THE SATURATED HYDROCARBONS

Methane (CH_4) is the first member of a series of hydrocarbons called the methane, paraffin, or alkane series. All are open chain linear compounds with the same characteristic chemical structure, similar chemical properties, and a regular gradation in physical properties. Thus a knowledge of the chemical and physical properties of a few representative members is sufficient to give an adequate working knowledge of the entire series. The more commonly encountered members of the saturated hydrocarbon family, together with pertinent physical properties, are listed in table 18-1, page 94. A series of similar compounds, such as those shown in the table, is called a *homologous series.* The formula of each compound in the homologous series of saturated hydrocarbons differs from that of the compound adjacent to it by the increment CH_2. The general formula for any member of the series is C_nH_{2n+2} where *n* represents the number of carbon atoms.

Chemical Reactivity

The chemical reactivity of the saturated hydrocarbons is typified by that of gasoline. Hazardous properties depend primarily on flash point and flammable limits. The saturated hydrocarbons are rather unreactive chemically. At room temperature, they are inert

Name	Empirical and Structural Formula	BP °F	Flash Point °F	Flammable Limit % by Volume in Air	
				Upper	Lower
methane	CH_4	-259	gas	14	5.3
ethane	C_2H_6 (CH_3-CH_3)	-128	gas	12.5	3.0
propane	C_3H_8 (CH_3-CH_2-CH_3)	-44	gas	9.5	2.2
butane	C_4H_{10} (CH_3-CH_2-CH_2-CH_3)	31	gas	8.5	1.9
pentane	C_5H_{12} (CH_3-CH_2-CH_2-CH_2-CH_3)	97	-40	7.8	1.5
hexane	C_6H_{14} (CH_3-CH_2-CH_2-CH_2-CH_2-CH_3)	156	-7	7.5	1.2
heptane	C_7H_{16} (CH_3-CH_2-CH_2-CH_2-CH_2-CH_2-CH_3)	209	25	6.7	1.2
octane	C_8H_{18}	258	56	3.2	1.0
nonane	C_9H_{20}	303	88	2.9	0.8
decane	$C_{10}H_{22}$	345	115	2.4	0.8
cyclohexane	C_6H_{12}	178	12	7.75	1.26

Table 18-1. Saturated Hydrocarbons.

with acids, bases, oxidizing agents, and reducing agents. The halogens (fluorine, chlorine, bromine, and iodine) react with the saturated hydrocarbons even at room temperature, replacing the hydrogen atoms in stepwise fashion. Oxygen reacts with the saturated hydrocarbons at slightly elevated temperatures.

THE UNSATURATED HYDROCARBONS

Another series of hydrocarbons, called the olefin or alkene series, has two hydrogen atoms less than the number required to satisfy the valence of carbon. This is best illustrated by comparing ethane (saturated) and ethylene (unsaturated).

ethane	ethylene
H H	H H
H:C:C:H	C::C
H H	H H

All members of the olefin series derive from the corresponding saturated compound by removing hydrogen atoms from adjacent carbon atoms:

$$CH_3 - CH_3 \rightarrow CH_2 = CH_2 + H_2$$
$$ethane \qquad ethylene$$

The ending *ane* on ethane signifies a saturated hydrocarbon. Similarly *ene* indicates an olefinic unsaturated hydrocarbon. The general formula for the olefin series is C_nH_{2n}. Some examples of the common olefinic hydrocarbons are listed in table 18-2.

Chemical Reactivity

In contrast to the saturated hydrocarbons, the olefins are highly reactive. The double bond is easily converted to a saturated bond by addition of other molecules.

$$CH_2 = CH_2 + Br_2 \rightarrow CH_2 - CH_2$$
$$Br \quad Br$$
$$CH_2 = CH_2 + H_2 \rightarrow CH_3 - CH_3$$
$$CH_2 = CH_2 + HOH \rightarrow CH_3CH_2OH$$
$$CH_2 = CH_2 + CH_2 = CH_2 \rightarrow CH_3 - CH_2 - CH = CH_2$$

Needless to say the olefins are highly flammable with oxygen. In general, they have much lower boiling points than the correspond-

Name	Empirical and Structural Formula	BP °F	Flash Point °F	Flammable Limit % Volume in Air	
				Upper	Lower
ethylene	C_2H_4 $CH_2{=}CH_2$	−153.4	gas	28.60	2.75
propylene	C_3H_6 $CH_3CH{=}CH_2$	−53	gas	11.10	2.00
butylene	C_4H_8 $CH_3CH_2CH{=}CH_2$	−23	gas	9.95	1.65
pentene	C_5H_{10} $CH_3CH_2CH_2CH{=}CH_2$	−53	gas	8.70	1.42

Table 18-2. Olefinic Hydrocarbons.

ing members of the saturated series. They tend to be more hazardous because of the lower flash point.

THE ACETYLENE SERIES

The acetylene series is even more unsaturated than the olefin series. This series of hydrocarbons can be thought of as resulting from the loss of two hydrogen atoms from adjacent carbon atoms. The general formula is C_nH_{2n-2}.

$$CH_3 - CH_3 \rightarrow CH{\equiv}CH + 2H_2$$

$$\text{ethane} \qquad \text{acetylene}$$

They are quite similar to the corresponding saturated and olefinic hydrocarbons in physical properties. Being much more highly unsaturated, they are correspondingly more reactive chemically. The nature of the reactions is similar to those of the olefins.

The only acetylenic compound commonly encountered is acetylene itself, primarily because of its extensive use in the oxyacetylene torch. It is also an inexpensive raw material used in a wide variety of chemical manufacturing processes. The fire fighting hazards associated with acetylene are discussed in unit 14.

THE AROMATIC HYDROCARBONS

The aromatic hydrocarbons constitute another major group. Benzene can be thought of as the parent compound in the aromatic hydrocarbon series much as ethylene as the parent compound in the olefinic series and acetylene in the acetylenic series. The benzene molecule is a symmetrical hexagonal ring consisting of six carbon atoms and six hydrogen atoms. A great deal of confusion results from the fact that the molecular formula can be written in two equivalent forms.

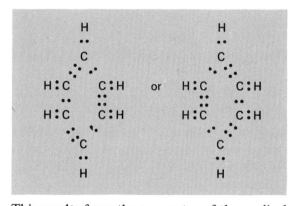

This results from the symmetry of the cyclical six carbon ring. If the carbon atoms are pictured as being stationary, the electron octets required for chemical bonding can be achieved in either of the two ways shown. In actual practice, electron pairs shift back and forth between the two structures. This is called *resonance*. The properties of the aromatic series of compounds are those conferred by this resonance.

Electron pairs in benzene can be represented by lines drawn between the atoms as has been done previously.

Electron dot	Structural	Name
Na **:** Cl	Na — Cl	sodium chloride
H **:** C **:** H (with H above and below)	H—C—H (with H above and below)	methane
O **::** C **::** O	O=C=O	carbon dioxide
benzene ring (electron dot)	benzene ring (structural)	benzene

Hazards

Most aromatic hydrocarbons are flammable. The comparative hazard is determined by flash point and flammable limits. Burning aromatic hydrocarbons can often be identified by unusually smoky flames.

The toxicity hazard of the aromatic hydrocarbons is generally underestimated — considering their extensive use. They are cumulative poisons, damaging to the liver and causing cancer, or leukemia. Even short-term exposure is dangerous because of their anesthetic effect.

For the sake of convenience, the benzene ring symbol is usually simplified by omitting the symbols for carbon and hydrogen. This is the most commonly used convention and is used in this text.

MISCELLANEOUS PETROLEUM PRODUCTS

A wide variety of hydrocarbons, largely products of the petroleum industry, are encountered which are potentially hazardous. Properties of these petroleum fractions are listed in table 18-4. For the most part, these materials are not pure chemical compounds for which a structural formula can be given. They are largely distillation fractions with known boiling point ranges or modified materials prepared from them. The product designations are those of the American Petroleum Industry (API) and are used universally within the industry.

Some of the common aromatic hydrocarbons are listed in table 18-3. The large listing is a reflection of the importance of aromatic compounds in modern technology.

Name	Formula	BP °F	Flash Point °F	Flammable Limit % by Volume in Air	
				Upper	Lower
benzene	C_6H_6	176	12	7.10	1.40
toluene	C_7H_8	231	40	6.75	1.27
xylene					
ortho	C_8H_{10}	292	63	6.00	1.00
meta	C_8H_{10}	282	77	7.00	1.10
para	C_8H_{10}	281	77	7.00	1.10
naphthalene	$C_{10}H_8$	424	174	5.90	1.00
anthracene	$C_{14}H_{10}$	671	250	- - - -	- - - -

Table 18-3. **Aromatic Hydrocarbons.**

Name	BP °F	Flash Point °F
gasoline	100 - 400	–35 to –50
petroleum ether, ligroin	100 - 140	–40
naphtha	200 - 350	30 - 100
kerosene	350	105 - 130
fuel oil	100 - 150	300 - 600
lubricating oil	700	400
mineral oil	700	400
greases	- - -	high
paraffin wax	- - -	high
asphalt	- - -	50 - 400
crude petroleum	- - -	low

Table 18-4. **Hazardous Properties of Petroleum Products.**

FLAMMABLE LIQUIDS

The hydrocarbons that fire fighters come in contact with most often are in the lower paraffin series, gasoline and fuel oils. These are classified according to their flash points. There are three classifications of flammable and combustible liquid as defined by the NFPA; these were shown in table 10-1.

Gasoline, with a flash point between –35 to –50°F, is defined as a flammable Class 1A liquid, while fuel oil #1 (kerosene) is defined as a combustible Class 2 liquid, since it has a flash point of approximately 115°F.

The behavior of these two liquids differs only by a factor of temperature. For instance, if the ambient temperature in an area is 80°F, and there is a spill of gasoline and one of kerosene, the immediate threat of ignition comes from the gasoline, since at 80°F the gasoline is giving off vapors suitable for ignition. No vapors are given off by the kerosene unless a temperature of 115°F is reached. If the same spill were in an area where the temperature was above 115°F, then the hazard risk would be the same for both the gasoline and kerosene.

FACTORS WHICH DETERMINE EMERGENCY ACTION

In handling any type of emergency with hydrocarbons, there are certain factors concerning the hydrocarbon the fire fighter must keep in mind: identification, flashpoint and ambient temperature, flammable limits, specific gravity, water miscibility, vapor density, and ignition temperature. These factors, all defined earlier, must be considered before corrective action is initiated.

• Identification – What is the liquid?

• Flash point and ambient temperature – Is the ambient temperature higher than the listed flash point for the liquid or is it lower? If the temperature is higher, then the fire fighter knows the liquid is giving off flammable vapors. On the other hand, if the flash point is higher than the ambient temperature, then vapors are not being given off. Fire fighting methods depend heavily on this information.

• Flammable limits – What are the limits? Are they very narrow (1 to 3) or are they wide (3-14, or more)? A greater chance of the vapors igniting is indicated with the wider range than with the narrower range.

• Specific gravity – What is the weight of liquid compared to water (water is 1.0)? Is the specific gravity greater than 1.0? Greater than 1.0 indicates that the liquid is heavier than water and, therefore, the water would float on the top. If the specific gravity is less than 1.0 as is the case with most hydrocarbons, then the hydrocarbon floats on top of the water, thus limiting the effectiveness of water as an extinguishing agent.

• Water miscibility – Is the liquid soluble in water? If so, to what extent? The

97

flash point of a water soluble liquid decreases as water is added. Infinite dilution decreases the chance of a fire or explosion hazard.

- Vapor density — What is the weight of the vapors given off, compared to the weight of air (vapor density of 1.0)? Will the vapors rise (vapor density less than 1.0) or will they seek lower ground (vapor density greater than 1.0)? Fire fighting operations and placement of apparatus depend on this important information.

- Ignition temperature and ignition source — The general practice here is to assume that if an ignition source is present, it is sufficient to cause ignition of any vapors present. Ignition sources (matches, sparks) are generally about 1500°F or higher. The ignition temperatures of the majority of hydrocarbons are lower than 1000°F. Other factors such as wind direction and humidity must not be overlooked when determining a course of action to be taken.

FIGHTING HYDROCARBON FIRES

There are some general principles to remember when fighting hydrocarbon fires. These principles apply whether the fire fighter is engaged in trying to control a tank truck fire on an interstate highway or a fire in a tank farm, where stationary tanks are present.

One of the greatest hazards in this situation is that of excessive pressure buildup and subsequent rupture of the tank. The vapor area of the tank must be kept cool at all times. Cooling prevents the excessive pressure buildup which can result in an explosion (BLEVE) if the pressure relief valve(s) is unable to relieve the pressure. Fire fighters should cool the tank immediately upon arriving; this can be done in any manner in which the department

is set up for initial attack. The number of lines needed depends upon conditions such as the type of tank.

As the pressure inside the tank is relieved through the pressure relief system, a loud whistling noise will be heard. If this noise increases while water is being applied, it is an indication that the pressure is still building up, that enough water is not being applied to effectively cool the tank. If an increase in water does not reduce the increase in pressure, or if additional water supply is not available, then unmanned monitors or hose holders should be used and the area evacuated.

At no time should any individual or apparatus be positioned at the ends of horizontal tanks. These ends of the typical shells are the weakest points and most likely to rupture if internal pressure becomes too great.

The rupturing of a tank is less likely to happen with tank trucks constructed of aluminum. The aluminum will melt (vapor area only) from direct flame impingement when temperatures in excess of 1000°F are reached. Although the chances for rupture are then decreased, opening of the tank allows the hydrocarbon to spill out onto the surface, thus creating a larger spill and increasing the fire hazard.

Some stationary horizontal tanks present the additional problem of their supports weakening from heat fatigue. If the supports collapse, the tank falls, causing a "flood" of flammable liquid. In addition to cooling the vapor area in this type of fire, the supports must be cooled, and the fire swept from under the tank.

Most departments carry foam or light water on their vehicles or have "special call" foam units available. These units are especially useful in the containment of hydrocarbon fires.

Preplanning is essential in stationary tank farms. Knowledge of the location of the power supply, shut off valves, and hydrants as well as of the slope of the ground, is of extreme importance in attacking and controlling any type of emergency.

REVIEW

1. The ending "ane" on a hydrocarbon indicates that it is

   a. Linear.
   b. Branched.
   c. Unsaturated.
   d. Saturated.

2. Which of the following compounds is a saturated hydrocarbon?

   a. Cyclohexane.
   b. Benzene.
   c. Ethylene.
   d. Acetylene.

3. The saturated hydrocarbons are chemically reactive with

   a. Oxygen at room temperature.
   b. Chlorine at room temperature.
   c. Reducing agents at room temperature.
   d. Acids at room temperature.

4. Which of the following hydrocarbons is the most reactive chemically?

   a. Cyclohexane.
   b. Propane.
   c. Ethylene.
   d. Acetylene.

5. Which of the following hydrocarbons is most apt to produce a smoky flame?

   a. Naphthalene.
   b. Nonane.
   c. Butylene.
   d. Methane.

6. Natural gas is largely composed of

   a. Decane.
   b. Hexane.
   c. Heptane.
   d. Methane.

7. The hazardous flammable properties of ligroin are most similar to those of

   a. Naphtha.
   b. Benzene.
   c. Gasoline.
   d. Ethylene.

8. The flash point of crude oil most closely resembles that of

   a. Paraffin wax.
   b. Mineral oil.
   c. Fuel oil.
   d. Naphtha.

Unit 19 Substituted Hydrocarbons

All other chemical compounds derived from the hydrocarbons are formed by the substitution of other atoms or groups of atoms for the hydrogen atoms in one of the hydrocarbons. Thus, substitution of a chlorine atom for a hydrogen atom forms the chlorinated series of the halocarbons. Substitution of a hydroxyl group (OH as in H-OH) for a hydrogen forms the alcohols. These and the other major groups of substituted hydrocarbons are shown in table 19-1. The nature of the substituting group determines the general physical and chemical properties of the new homologous series of compounds. All of the compounds listed in table 19-1 are basic industrial compounds used in large quantities.

ALCOHOLS

All of the common alcohols listed in table 19-2 are low flash point materials and are therefore flammable hazardous materials.

Some of the properties characteristic of alcohols are listed: (1) The lower molecular weight alcohols burn with lower flame temperatures because they are already partially oxidized. As the molecular weight increases the flame temperature of the parent hydrocarbon is approached; (2) the lower molecular weight alcohols are water soluble; those with five or more carbon atoms become progressively more water insoluble as their hydrocarbon component predominates; (3) all alcohols except ethyl alcohol are highly toxic; (4) the hazards associated with alcohol fires are those of any low flash point material; (5) all alcohols have specific gravities less than 1; (6) all alcohols have vapor densities greater than 1.

Of the many alcohols known in the laboratory, only a few are used in large quantity. This is a direct result of modern technology and centers around the availability of cheap industrial materials. Most *ethyl alcohol* is

Class of Compound	Characteristic Group	Formula	Name
alcohol	OH	CH_3CH_2OH	ethyl alcohol
amine	NH_2	$CH_3CH_2NH_2$	ethyl amine
fluoride	F	CH_3CH_2F	ethyl fluoride
chloride	Cl	CH_3CH_2Cl	ethyl chloride
bromide	Br	CH_3CH_2Br	ethyl bromide
iodide	I	CH_3CH_2I	ethyl iodide
aldehyde	CHO	CH_3CHO	acetaldehyde
acid	COOH	CH_3COOH	acetic acid
amide	$CONH_2$	$CH_3CH_2CONH_2$	ethyl amide
ester	$COOCH_2CH_3$	$CH_3COOCH_2CH_3$	ethyl acetate
ether	O	$CH_3CH_2OCH_2CH_3$	diethyl ether
ketone	C = O	$CH_3CH_2\underset{\underset{O}{\|\|}}{C}CH_2CH_3$	diethyl ketone

Table 19-1. Formation of Substituted Hydrocarbons From Ethane (CH_3-CH_3).

Name and Formula	BP °F	Flash Point °F	Occurrence or Use
methyl alcohol CH_3OH	147	52	Fuel, solvent for shellac, starting material for manufacture of formaldehyde
ethyl alcohol CH_3CH_2OH	173	55	Fuel, beverages, starting material for manufacture of ether, chloroform, acetic acid, dyes, lacquers, varnishes
propyl alcohol $CH_3CH_2CH_2OH$	207	77	Mainly laboratory use
isopropyl alcohol $CH_3\text{-}CH\text{-}CH_3$ $\quad\quad OH$	181	53	Used in manufacture of isopropyl acetate which in turn is a lacquer solvent. Isopropyl alcohol is sold as rubbing alcohol.
butyl alcohol $CH_3CH_2CH_2CH_2OH$	243	84	Mainly laboratory use
secondary butyl alcohol $CH_3CH_2CH\text{-}CH_3$ $\quad\quad\quad OH$	201	75	Used in manufacture of butyl acetate which in turn is a lacquer solvent

Table 19-2. Properties of the Alcohols.

made by addition of water to ethylene, a flammable gas, which is a major by-product of the petro-chemical industry. *Isopropyl alcohol* is produced by addition of water to propene, another major by-product of the petrochemical industry. *Propyl alcohol* is not a commonly encountered alcohol. *Isobutyl alcohol* is made by the addition of water to butene, another major by-product of the petrochemical industry. Isobutyl alcohol rather than *butyl alcohol* is formed in the reaction and is therefore the commonly encountered form.

Glycols

Glycols are alcohols containing two or more hydroxyl groups. They are often referred to as the higher alcohols. Examples of some of the more common glycols are glycerol, ethylene glycol, and propylene glycol. The properties of the glycols are summarized in table 19-3, page 102.

The glycols all have high flash points, and are, therefore, not particularly flammable hazardous materials although they are definitely combustible. Characteristically glycols are generally nontoxic unless ingested, have a specific gravity and a vapor density greater than 1, and are water soluble.

Fire Fighting Procedures

The procedure for fighting a fire involving alcohols is the same as other flammable and combustible liquids. One important factor must be kept in mind, however: alcohol and glycols are not compatible with all foams. They degrade the foam and break it down, thus limiting its usefulness. This is due to the fact that most alcohols are water soluble, in contrast to other flammable liquids such as gasoline, which is not water soluble. It is this interaction between the alcohol, water and foam that limits its effectiveness.

Name and Formula	BP, °F	Flash Point, °F	Occurrence or Use
ethylene glycol CH$_2$-CH$_2$ $\mid$ $\mid$ OH OH	387	232	Permanent antifreeze

propylene glycol
CH$_3$-CH-CH$_2$
    $\mid$ $\mid$
    OH OH
	370	210	Used in manufacture of explosives and plastics

glycerol CH$_2$-CH-CH$_2$ $\mid$ $\mid$ $\mid$ OH OH OH	554	320	Used in manufacture of explosives
diethylene glycol HOCH$_2$CH$_2$OCH$_2$CH$_2$OH	480	290	Corrosion inhibitor

Table 19-3. Properties of the Glycols.

Alcohol-type foams are a low protein type which have been formulated to give a greater breakdown resistance when used with water soluble fuels such as alcohols. These types of foams are used in proportions of 6 percent or more. Alcohol-type foam should not be used with other foams unless the two are listed as compatible; mixed use of two incompatible foams limits their effectiveness.

AMINES

Amines may be viewed as being derivatives of ammonia. The more common amines are summarized in table 19-4.

Name and Formula	BP, °F	Flash Point, °F	Occurrence or Use
ammonia NH$_3$	-27	gas	Basic chemical in manufacturing processes, cleaning compounds
methylamine CH$_3$NH$_2$	21	gas	Starting material for plastics manufacturing, solvents, pharmaceuticals
dimethylamine (CH$_3$)$_2$NH	45	gas	"
trimethylamine (CH$_3$)$_3$N	38	gas	"
ethylamine CH$_3$CH$_2$NH$_2$	62	0	"
propylamine CH$_3$CH$_2$CH$_2$NH$_2$	120	-35	"
butylamine CH$_3$CH$_2$CH$_2$CH$_2$NH$_2$	172	10	"
ethylenediamine CH$_2$ - CH$_2$ $\mid$ $\mid$ NH$_2$ NH$_2$	241	110	"
diethylenetriamine CH$_2$-CH$_2$-N-CH$_2$-CH$_2$ $\mid$ $\mid$ $\mid$ NH$_2$ H NH$_2$	404	215	"

Table 19-4. Properties of the Amines.

Amines all have characteristic hazardous properties. They are highly toxic, have a low flash point, a specific gravity less than 1, and a vapor density greater than 1 (except for ammonia). They produce the toxic combustion products (HCN and NH_3) and are caustic (because they are strong bases). This aspect is discussed in greater detail in later units.

Since the amines are so hazardous, it is fortunate that they are easily detectable because of their odor. Most have the characteristic pungent odor of ammonia. All have such disagreeable odors that voluntary exposure is highly unusual. The toxicity of the amines warrants comment at this point. They are toxic by ingestion, inhalation, and skin absorption. Because of the skin absorption characteristic, even gas masks or positive breathing apparatus do not provide complete protection.

The common amines are those which can be manufactured inexpensively from ammonia methane, ethylene, propylene, and butylene. Other amines are simply not manufactured in large quantity even though they may have special desirable properties.

Fire Fighting Procedures

The hazard of fire fighting in association with amines is not only flammability, but also the threat to health. The majority of amine-containing compounds are highly toxic. Some of the organic amines are classified as poisons and are required to be labeled as such. Water, dry chemical foam, or carbon dioxide can be used in extinguishment of these types of fires.

HALOCARBONS

The halocarbons are formed by the substitution of a halogen (fluorine, chlorine, bromine, or iodine) for the hydrogen of a hydrocarbon. Examples of the more common halocarbons are listed in table 19-5.

Name and Formula	BP, °F	Flash Point, °F	Occurrence or Use
chloromethane CH_3Cl	-11	- - -	Solvent
carbon tetrachloride CCl_4	170	- - -	Cleaning solvent
chloroform $CHCl_3$	142	- - -	Anesthetic
chlorobenzene C_6H_5Cl	270	- - -	Solvent
bromobenzene C_6H_5Br	311	- - -	Solvent
paradichlorobenzene $C_6H_4Cl_2$	343	- - -	Moth crystals
methylene chloride CH_2Cl_2	104	- - -	Solvent
perchlorethylene C_2Cl_4	250	- - -	Degreaser
ethyl chloride C_2H_5Cl	53	- - -	Anesthetic
methyl bromide CH_3Br	41	- - -	Fumigant

Table 19-5. Properties of the Halocarbons.

Name and Formula	BP, °F	Flash Point, °F	Occurrence or Use
formaldehyde $HCHO$	–3	gas	Fumigant
formalin 37%$HCHO$ in H_2O	214	185	Biological preservant, embalming fluid
paraldehyde $C_3H_6O_3$	255	96	Plastics and leather manufacture
acetaldehyde CH_3CHO	70	–36	Chemical manufacturing
propionaldehyde CH_3CH_2CHO	120	15	Chemical manufacturing
benzaldehyde C_6H_5CHO	355	148	Chemical manufacturing

Table 19-6. **Properties of Aldehydes.**

The halocarbons have a low flammable hazard. Their specific gravity and vapor density are greater than 1. They are insoluble in water. They present a high toxicity hazard and produce toxic combustion products: chlorine compounds (HCl and $COCl_2$), bromine compounds (HBr and $COBr_2$), and fluorine compounds (HF and COF_2).

The hazards of the halocarbons are almost entirely toxicity and toxic combustion products. This aspect is discussed in greater detail later.

ALDEHYDES AND KETONES

The aldehydes and ketones are hydrocarbons in which a doubly bonded oxygen (=O) has replaced two hydrogen atoms. In *aldehydes*, this replacement is at an end carbon in the chain. In propionaldehyde ($CH_3CH_2\overset{\text{O}}{\underset{\text{||}}{C}}$ - H), two hydrogens on the end carbon atom of propane ($CH_3CH_2CH_3$) have been replaced by a doubly bonded oxygen.

If the substitution of the doubly bonded oxygen occurs at any carbon atom other than the end carbon, the product is a *ketone*. Thus, acetone (CH_3 - $\overset{\text{O}}{\underset{\text{||}}{C}}$ - CH_3) is also formed from propane ($CH_3CH_2CH_3$).

Commercially all aldehydes are manufactured by mild oxidation of alcohols. Thus, the commonly encountered aldehydes are those for which the parent alcohols are available in large quantity. Similarly, all ketones are made by catalytic oxidation of alcohols with the hydroxyl group on other than the end carbon atom.

The common aldehydes are shown in table 19-6 and the ketones, in table 19-7.

Name and Formula	BP, °F	Flash Point, °F	Occurrence or Use
acetone $CH_3\underset{\text{O}}{\overset{\text{||}}{C}}CH_3$	134	0	Solvent
methylethyl ketone $CH_3CH_2\underset{\text{O}}{\overset{\text{||}}{C}}CH_3$	176	21	Solvent

Table 19-7. **Properties of Ketones.**

The characteristic properties of the aldehydes are (1) low flash point (except for benzaldehyde), (2) properties such as vapor density, specific gravity, and water solubility are variable, (3) highly reactive chemically, and (4) moderately toxic. The corresponding properties of the ketones are (1) low flash point, (2) higher water solubility than corresponding aldehydes, (3) low chemical reactivity, and (4) low toxicity.

The aldehydes and ketones present similar degrees of flammable hazard since they are low flash point materials. The aldehydes are considerably more hazardous because of their higher toxicity and chemical reactivity.

Fire Fighting Procedures

The aldehydes and ketones are very similar to the alcohols in their hazard characteristics. Alcohol resistant foams should be used when fighting fires with these chemicals. The lower molecular weight liquids such as acetaldehyde and acetone are extremely flammable. Acetaldehyde and some of the other aldehydes oxidize in air and can form unstable peroxides that explode readily. Acetaldehyde can also polymerize, with the evolution of heat. The result of such a polymerization is a violent release of energy, causing great destruction.

ACIDS

The acids may be regarded as hydrocarbons in which all three hydrogen atoms on an end carbon are replaced by a doubly bonded oxygen ($=O$) and a hydroxyl group ($-OH$). The carboxylic acids of this type are very weak acids and none of their hazards are associated with acidity.

$$CH_3CH_2CH_3 \qquad\qquad CH_3CH_2\overset{\displaystyle O}{\overset{\displaystyle \|}{C}}\text{-}OH$$

propane $\qquad\qquad\qquad$ propionic acid

Some of the characteristic properties of carboxylic acids are high flash point, high chemical reactivity, and low corrosive hazard.

The organic acids do not offer such a high degree of hazard as the mineral acids, such as sulfuric acid, do.

Fire Fighting Procedures

The fire fighting procedures for the organic acids are the same as for the alcohols, aldehydes, and ketones.

ESTERS

Organic acids and alcohols react slowly with the elimination of a water molecule to form an ester. Acetic acid and ethyl alcohol, for example, react to form the ester called ethyl acetate. Similarly, butyl acetate is formed by the reaction of acetic acid and butyl alcohol.

$$CH_3\overset{\displaystyle O}{\overset{\displaystyle \|}{C}}\text{-}OH + CH_3CH_2OH \rightarrow CH_3\overset{\displaystyle O}{\overset{\displaystyle \|}{C}}\text{-}OCH_2CH_3$$

acetic acid $\quad$ ethyl alcohol $\qquad$ ethylacetate

$$CH_3\overset{\displaystyle O}{\overset{\displaystyle \|}{C}}\text{-}OH + CH_3CH_2CH_2CH_2OH \rightarrow CH_3\overset{\displaystyle O}{\overset{\displaystyle \|}{C}}\text{-}OCH_2CH_2CH_2CH_3$$

acetic acid $\quad$ butylalcohol $\qquad\qquad$ butyl acetate

The lower molecular weight esters are materials with pleasant odors and flavors and for that reason are extensively used in perfumes and food flavorings. It is apparent that the esters as a group are not toxic. Esters are encountered as solvents (which are highly flammable) and also as plasticizers in plastics. The properties of the more common esters are shown in table 19-8, page 106.

Fire Fighting Procedures

Esters, despite their pleasant odor, are highly flammable and the same precautions should be taken as for alcohols.

AMIDES

Amides are formed by the reaction of ammonia with organic acids. The mode of formation is quite similar to that of the esters.

Name and Formula	BP, °F	Flash Point, °F	Occurrence or Use
methyl acetate CH_3COOCH_3	140	14	Solvent
ethyl acetate $CH_3COOC_2H_5$	171	24	Lacquer solvent
propyl acetate $CH_3COOC_3H_7$	215	58	Lacquer solvent
butyl acetate $CH_3COOC_4H_9$	260	72	Lacquer solvent
amyl acetate $CH_3COOC_5H_{11}$	290	77	Solvent

Table 19-8. Properties of the Esters.

$$CH_3\overset{O}{\overset{||}{C}}\text{-OH} + NH_3 \rightarrow CH_3\overset{O}{\overset{||}{C}}\text{-NH}_2 + H_2O$$

acetic acid   acetamide

The most commonly encountered examples of amides are the proteins and nylon. Both are polymers and are discussed in greater detail in the unit on plastics. From a chemical point of view, the amides are hazardous primarily because of the formation of toxic combustion products, such as HCN. Some of the amides are inherently fire-retardant and have found use in protective turnout gear for fire fighters.

ETHERS

When two molecules of alcohol combine with the loss of one molecule of water an ether is formed. Diethyl ether is formed from two molecules of ethyl alcohol.

$$CH_3CH_2OH + CH_3CH_2OH \rightarrow CH_3CH_2\text{-O-}CH_2CH_3 + H_2O$$

diethyl ether

The ethers are highly hazardous materials. Some properties of the common ethers are listed in table 19-9. Other properties characteristic of the ethers are that they (1) are highly flammable because of low flash point, (2) have a specific gravity less than 1 and a vapor density greater than 1, (3) are hazardous because of anesthetic properties, and (4) are potentially explosive because of the formation of peroxides with the air. Dioxane is the only toxic ether commonly encountered.

Fire Fighting Procedures

Ethers have vapor densities which are considerably heavier than air. This important fact should be remembered when fighting ether fires or investigating any type of ether leak. Massive fires should be fought from "safe" distances, determined by prevailing fireground conditions.

Dry chemical, foam or carbon dioxide are suitable for extinguishment. Water may be ineffective because of the low flash point; however, water can be used to disperse unignited vapors.

Name and Formula	BP, °F	Flash Point, °F	Occurrence or Use
dimethyl ether $CH_3\text{-O-}CH_3$	–11	gas	Solvent
diethyl ether $CH_3CH_2\text{-O-}CH_2CH_3$	95	–49	Anesthetic, solvent
divinyl ether $CH_2{=}CH\text{-O-}CH{=}CH_2$	102	–22	Solvent

Table 19-9. Properties of Ethers.

INDUSTRIAL MATERIALS

A multitude of industrial materials are encountered which are potentially hazardous under certain circumstances. The use and hazardous properties of many of these materials (paints, varnishes, lacquers, insecticides, fungicides, weed killers, and cosmetics, to mention a few) are outlined in table 19-10.

Name	Use	Hazardous Properties
acetic anhydride	In manufacture of rayon and alkyd resins.	Corrosive because of reaction with water to form acetic acid. Highly toxic.
acrolein	Widely used industrial solvent.	Highly toxic.
aniline	Widely used chemical.	Intermediate toxicity characteristic of amines. Flash point 158°F.
butyl carbitol (diethylene glycol monobutyl ether)	Solvent.	Highly flammable. Flash point 240°F.
butyl cellosolve (ethylene glycol monobutyl ether)	Solvent.	Highly flammable. Flash point 165°F.
camphor (Solid hydrocarbon derived from laurel tree)	Similar to turpentine.	Hazardous flammable solid. Flash point 175°F.
dibutyl carbitol (diethylene glycol dibutyl ether)	Solvent.	Highly flammable. Flash point 245°F.
dibutyl cellosolve (ethylene glycol dibutyl ether)	Solvent.	Highly flammable. Flash point 185°F.
dioxane	Solvent extraction.	Flammable hazardous material. Flash point 65°F. Toxic.
fusel oil (isoamyl alcohol)	Solvent.	Highly toxic. Flash point 123°F.
varnish: Consists of a resin dissolved in a medium which largely evaporates. Other components are a drying oil, an oxidation catalyst commonly called a drier, and a thinner. Widely used solvents are turpentine, benzene, acetone, ethyl acetate, alcohols, and ethylene glycol monoethyl ether. Many types of resins are used but the most common are phenolformaldehyde, cellulose acetate, cellulose nitrate, and the alkyd resins.	Provides a protective coating.	Primarily flammable hazard. Toxicity is a minor hazard.
lacquer: Low flash point solvents such as acetone and the esters are used to provide rapid evaporation. Low resin contents are used to provide the desired thin film. Resins used are nitrocellulose, ethyl cellulose, cellulose acetate, and cellulose butyrate.	Provides a thin protective coating which dries quickly.	Hazardous flammable material. Negligible toxicity hazard.
shellac: Naturally occurring resin "lac" dissolved in ethyl alcohol.	Provides protective coating.	Hazardous flammable material.
turpentine: Liquid hydrocarbon derived from pine pitch.	Often used as a solvent in paints, varnishes, lacquers, and wax polishes.	Hazardous flammable material. Flash point 100°F.

Table 19-10. Composition and Hazardous Properties of Miscellaneous Industrial Materials.

REVIEW

1. The hazardous flammable properties of alcohols are those of a

   a. Water soluble material.   c. Flammable liquid.
   b. Low flash point material.   d. All of these.

2. A glycol is

   a. A type of alcohol.
   b. A low flash point material.
   c. Highly toxic because it contains two hydroxyl groups.
   d. Insoluble in water.

3. The hazardous property of amines is

   a. Low flash point.   c. Corrosivity.
   b. High toxicity.   d. All of these.

4. Amines produce the following toxic combustion products

   a. HCl and $COCl_2$.   c. N_2 and NO.
   b. NH_3 and HCN.   d. All of these.

5. An example of a ketone is

   a. Formalin.   c. Benzaldehyde.
   b. Acetone.   d. Decane.

6. Organic acids are

   a. Highly corrosive.
   b. Highly reactive.
   c. Low flash point materials.
   d. Hazardous because of their toxic combustion products.

7. The hazardous properties of amides are most similar to those of

   a. Acids.   c. Ketones.
   b. Aldehydes.   d. Amines.

8. Ethers are highly hazardous because

   a. They have low flash points.   c. Of high vapor density.
   b. They form peroxides with air.   d. All of these.

9. The only common toxic ether is

   a. Dioxane.   c. Diethyl ether.
   b. Dimethyl ether.   d. None of these.

10. Dioxane is an

    a. Ester.   c. Acid.
    b. Ether.   d. Aldehyde.

Unit 20 Plastics

Most students are familiar with plastics, but few know much about their flammable and toxic properties. Plastics are chemical compounds in which a large number of identical or similar molecules are joined together in a single large molecule, called a *polymer*. The polymerization of ethylene to form polyethylene is a familiar example.

$$CH_2 = C(H)_2 + CH_2 = CH_2 \rightarrow CH_3\text{-}CH_2\text{-}CH = CH_2$$

$$CH_3\text{-}CH_2\text{-}CH = C(H)_2 + CH_2 = CH_2 \rightarrow CH_3\text{-}CH_2\text{-}CH_2\text{-}CH_2\text{-}CH = CH_2$$

This process of addition of ethylene continues until a chain of hundreds of thousands of CH_2 groups is formed. The product is polyethylene and its chemical structure is often designated as $(CH_2)_x$ where x indicates the number of CH_2 groups in the plastic molecule.

A long linear chain of this type is called a *thermoplastic*. The common thermoplastics together with their hazardous properties are listed in table 20-1. All thermoplastic materials have similar flammable hazardous properties. They melt when heated. This property characterizes a major aspect of their flammable hazardous properties. Since they melt when heated, their flammable hazardous properties are those of a flammable liquid rather than a flammable solid. At elevated temperatures they break down (dissociate) into the monomers from which they were formed. Some

$$(CH_2)_x \longrightarrow CH_2 = CH_2$$

Polyethylene polymer   polyethylene monomer

plastics which contain the elements N, Cl, Br and F burn with the formation of toxic combustion products.

The other major class of plastics is the *thermosetting* plastics. They differ from the thermoplastics only in that they are cross-

linked. A cross-linked plastic is bonded so strongly that it cannot melt when heated. This inability to melt is the basis for the word "thermosetting." All thermosetting plastics exhibit the same combustion behavior as wood. Wood burns as it does because it is a highly crosslinked cellulose polymer. It does not burn until it is heated to a temperature sufficiently high to drive off gaseous molecules.

The common thermosetting plastics together with their hazardous properties are listed in table 20-2, page 111. All thermosetting plastics are similar in their flammable hazardous properties. They do not melt when heated. They burn like wood.

The significance of this to the fire fighter is that linear polymers (thermoplastics) are generally a greater fire hazard than cross-linked polymers (thermosetting plastics) because they can melt and liquefy. The more highly cross-linked the polymers, the less the magnitude of the fire hazard.

MONOMERS – THE BASIC BUILDING BLOCKS OF PLASTICS

The basic "building block" of any polymer, whether it is a thermoset or thermoplastic, is called the monomer. This comes from the word mono, meaning "one." The function of this monomer is to join with another monomer of its own type, or one of a different type in a continuous joining together that gives a long chain called a polymer.

Monomers can be liquid, solid or gaseous. In these three stages, the monomer can be flammable or nonflammable; it can be toxic or nontoxic. However, when these monomers react, their character completely changes. Styrene, for example, is a flammable liquid that is toxic and has a flash point of 88°F.

Monomer	Flash Point of Monomer °F	Solid Flammable Hazard	Toxic Combustion Products
polyethylene $CH_2 = CH_2$	gas	high	- - -
polypropylene $CH_3CH = CH_2$	gas	high	- - -
polyisobutylene $CH_2 = C$ with two CH_3 groups	gas	high	- - -
polyvinylacetate $CH_2 = CHCOOCH_3$	18	high	- - -
polyvinyl chloride $CH_2 = CHCl$	gas	high	hydrochloric acid, phosgene HCl, $COCl_2$
polyvinylidene chloride $CH_2 = CCl_2$	5	high	HCl, $COCl_2$
polyvinylalcohol $CH_2 = CH$ – OH	85	high	- - -
polyacrylonitrile $CH_2 = CHCN$	32	high	hydrogen cyanide, ammonia HCN, NH_3
polyacrylamide $CH_2 = CH\ CONH_2$	- - -	high	HCN, NH_3
polyacrylic acid $CH_2 = CH\ COOH$	130	high	- - -
polystyrene $CH = CH_2$	88	high	- - -
polymethylmethacrylate $CH_2 = C$ (CH_3, $COOCH_3$)	50	high	- - -
polytetrafluoroethylene $CF_2 = CF_2$	- - -	low	HF, COF_2
polytrifluorochloroethylene $CF_2 = CFCl$	- - -	low	HF, HCl, COF_2, COC
Amine Base Plastics			
nylon adipic acid $HOOC(CH_2)_4COOH$	- - -	high	HCN, NH_3
hexamethylene diamine $H_2N(CH_2)_6NH_2$	- - -		
polyvinylcarbazole acetylene $CH \equiv CH$	gas	high	HCN, NH_3
carbazole	- - -		

Table 20-1. Hazardous Properties of Thermoplastics.

Monomer	Flash Point of Monomer °F	Solid Flammable Hazard	Toxic Combustion Products
Phenol-Formaldehyde Plastics			
phenol ⬡OH	- - -	moderate	- - -
formaldehyde HCHO	gas		- - -
Phenol-Furfural Plastics			
phenol ⬡OH	- - -	moderate	- - -
furfural (furan)CHO	140		
Melamine Plastics			
melamine (structure)	- - -		HCN, NH$_3$
formaldehyde HCHO	gas	moderate	
Urea Formaldehyde Plastics			
urea H$_2$N\ /C=O H$_2$N	- - -		HCN, NH$_3$
formaldehyde HCHO	gas	moderate	
Epoxy Plastics			
ethylene oxide CH$_2$-CH$_2$ \O/	gas	moderate	- - -
propylene oxide CH$_2$-CH-CH$_3$ \O/	gas	moderate	- - -
epichlorohydrin CH$_2$-CHCH$_2$Cl \O/	gas	moderate	HCl, COCl$_2$
Polyurethane Plastics			
Many types of formulations are marketed but all are typically formed from isocyanates and glycols.			
toluenediisocyanate CH$_3$ ⬡NCO NCO	- - -	moderate (except poly-urethane foam)	HCN, NH$_3$
ethylene glycol HO(CH$_2$)$_2$OH	- - -		
Polyester Plastics			
Many types of formulations are marketed but all are typically formed from organic acids and organic glycols.			
adipic acid HOOC(CH$_2$)$_4$COOH	- - -	moderate	- - -
ethylene glycol HO(CH$_2$)$_2$OH	- - -		
Silicone Plastics			
All silicone plastics tend to be nonflammable or only slightly flammable. They ordinarily do not have toxic combustion products. They are nonhazardous plastics.			

Table 20-2 Hazardous Properties of Thermosetting Plastics

This liquid when polymerized forms the versatile solid polymer polystyrene which has a multitude of uses.

POLYMERS

Polymers can be either liquid or a solid. They are also called resins. These resins can be made into plastics by molding into a special form, or fillers can be added and the resin extruded or cast. The type of resin used depends on the application.

Polymers can be produced by heating monomers together; they can be synthesized by reacting with ultra violet or infra red light or plain sunlight; or they can be catalyzed.

Hazards

Manufacturers have added catalysts to speed up reactions. (A *catalyst* is a chemical which speeds up a reaction but does not enter into the reaction itself.) The catalysts used are generally chemicals that contain their own oxygen supply, such as *peroxides*. These peroxides are highly flammable and extremely reactive solids and liquids. The fire fighter should be aware of plastic-producing plants in his district, the type of peroxide catalyst being used, and the amount stored at the plant. The highly reactive peroxide compounds should be stored in a cooled isolated building.

Catalysts other than peroxide may be found in plastic plants. Metal halides such as diethylaluminum chloride (DEAK) and metal alkyls such as triisopropyl-aluminum are used and are extremely flammable and sensitive to fire fighting operations. The metal alkyls react in air spontaneously and vigorously. Like sodium, they must be kept inert. The metal alkyls are kept in organic solvents such as toluene. The alkyl aluminums should not have water, foam, or halogenated agents used on them. Full protective gear should be worn

since the fumes from the burning aluminum alkyls are highly toxic. The metal halides, although safe from combustion, do give off hydrochloric and hydrofluoric acid when moist air or water reacts with them.

Plastics sometimes have an organic compound added to them in varying amounts to give the final product certain characteristics such as good flexibility or good flow. These compounds are called plasticizers. Plasticizers can either increase the flammability or can be of such a nature as to inhibit the combustion of the plastic.

Hazards of Plastics Use

For economic reasons plastic or synthetic materials make very attractive replacements for natural or metal products. Their substitution has caused some problems, however. Early substitution of synthetic materials was not done with the thought of flammability in mind. These earlier products were hazardous and added to the spread of fire and smoke.

Use in some nursing homes is an example of this situation. Fireproof and resistant structures were built — only to have the interiors equipped with materials containing synthetics. Synthetic carpets, drapes, and furniture when burned not only gave off toxic fumes, but also, as in the case of some types of carpeting, gave off considerable amounts of dense smoke. This smoke has been responsible for asphyxiation deaths for many elderly persons.

The plastics industry has partially alleviated this problem by incorporating halogen or phosphate compounds into their products. Both are effective in decreasing the combustibility of the synthetically produced materials.

These products do have drawbacks; toxic compounds such as phosgene and hydrochloric, hydrobromic, hydrofluoric and phosphoric acid are given off when these plastics are burned or heated to high temperatures.

ABS

The combination of acrylonitrile, butadiene and styrene monomers gives a very tough polymer called ABS. This type of polymer burns at a slow rate. Flame retardant varieties are available for home use. ABS is used in plastic pipe, telephones, automotive applications, appliances such as refrigerator interiors, and furniture.

ACRYLICS

Clarity and toughness are characteristics of the acrylics. Methyl methacrylate is the monomer used for its production. Because of its clarity, its biggest use is in the advertising industry for signs. Other uses are in the automotive areas for paints and lenses, in appliances, in packaging and in the building industry. Unless treated with flame retardants, the acrylics are essentially combustible.

CELLULOSE

Cellulose polymers are one of the few natural polymer systems in use today. Cellulose is a natural polymer that through chemical modification is made into such familiar polymers as cellulose acetate, cellulose nitrate, and ethyl cellulose.

Cellulose nitrate is a familiar thermoplastic that is highly flammable. Called *Celluloid,* this was the first commercial plastic in use. It was combined with a plasticizer, camphor, which is a flammable solid. This type of polymer was patented in the late 1800s and was commonly called pyroxylin plastic.

Cellulose nitrate is prepared by the reaction of cellulose with nitric acid. When processed, it is in the form of a fibrous, pulplike material or in the form of powder. If kept in the dry state, cellulose nitrate exhibits a high degree of flammability. Therefore, it is usually kept wet (20%-35%) with water or alcohol. The storage of this polymer should be in protected areas. Drums should be kept away from heat or any source which fosters drying out. In the dry state, the cellulose nitrate decomposes gradually. This slow decomposition releases gases which are not only toxic but flammable. Since cellulose nitrate contains a high percentage of nitrogen (11%) some of the toxic gases given off contain nitrogen oxides and hydrogen cyanide.

Fire Fighting Procedure

Fires involving cellulose nitrates should be fought with extreme caution. This polymer contains its own oxygen which makes it burn furiously. The use of unmanned monitors or hose holders is recommended for this type of fire. Self-contained breathing apparatus is also recommended because of the toxic gases given off.

Cellulose acetate polymers are prepared from the reaction of cellulose with acetic acid and acetic anhydride. Reacting the cellulose with butyric or propionic anhydride along with acetic acid gives polymers called cellulose acetate butyrate and cellulose propionate.

EPOXIES

Strength is the key feature of epoxies. They are a thermoset class of resins. They are used for binders, maintenance and marine coatings, electrical potting and encapsulating, paints, and adhesives.

Epoxies were introduced in the early 1950s and have expanded into several categories. The epoxies are two part component systems. The main part consists of the epoxy groups themselves, while the second part contains a catalyst. The catalysts are of different types. The fire fighter should be especially aware of one type, the amines. Amines are generally toxic compounds and present health hazards. Another catalyst that presents health hazards is boron trifluoride which is extremely irritating to the eyes and respiratory tract.

POLYESTERS

The family of polyesters presents a wide variety of products and uses to the consumer. Automotive bodies, carpets, draperies, furniture molding compounds, and clothing are all applications that use polyester resins.

Polyesters are made from a polyfunctional alcohol (ethylene glycol or glycerol) and a dibasic acid such as terephthalic, fumaric or maleic.

These resins are combustible and most frequently have flame retardants added to them or incorporated into the chemical background of the polymer for flame resistance.

One category of polyesters, called the unsaturated polyesters, are cured by the use of peroxide catalysts. Fire fighters should be aware of the manufacturing plants in their area that use polyesters and the highly reactive peroxides.

FLUOROPLASTICS

Polymer structures which contain fluorine in their backbone are called fluoroplastics. Teflon® and Kel-F® are familiar examples. Fluoroplastics are unique in that they are resistant chemically to corrosive liquids and most solvents. They are useful in electrical valves and gaskets and also where inherent flame retardancy is needed. They are also noted for their "nonstick" characteristics. Examples of common fluoroplastics are shown.

Abbreviation	Name
CTFE	Chlorotrifluoroethylene
E-CTFE	ethylene-chlorotrifluoroethylene
ETFE	ethylene-tetrafluoroethylene
PVF$_2$	polyvinylidene fluoride
PFA	perfluoroalkoxy
TFE	tetrafluoroethylene
FEP	fluorinated ethylene-propylene

NYLON

Nylon is a very familiar and common polymer which is prepared using an acid and an amine. This alphatic acid is usually adipic and the amine is generally hexamethylene diamine. However, through the years, new and improved copolymers of nylon have brought different monomers into the polymer backbone.

Nylon is known for its toughness and rigidity and is used in a variety of applications familiar to the student.

Flame retardants are added to the aliphatic type of nylons to render them stable to flame. Nylons using all aromatic acids have been given the term "aramid." These nylons are inherently flame retardant and have found use in protective turnout clothing for hazardous occupations such as fire fighting and race car driving.

PHENOLIC

Phenolic resins are thermoset resins and are the oldest manmade polymers. Phenolic resins are prepared from the dangerous acid phenol (carbolic acid) and formaldehyde. The resin can be found in the solid form, in water solution or in organic solvent. The greatest application is in the solid form where the resin is used as a molding compound.

Phenolics are found in a wide variety of applications, including insulation, laminates, resins for brake linings, some paints and adhesives. When heated, phenolics decompose instead of melting.

POLYAMIDE-IMIDES

Polyamide-imides are a class of polymers in the high temperature category. They are found mainly in organic solvent systems which are flammable. Amide-imides, in general, are inherently flame retardant, however. They are used in a solid form in molding applications.

In the liquid form, amide-imides are used for enamels for magnet wire manufacture. In the solid form, they are used as molding compounds in electronic components, in valves, pumps, and gears.

Hazard to health is a potential problem to the fire fighter with the monomers used in the manufacture of amide-imides. Amines or isocyanates are sometimes used and have been known to be toxic. The fire fighter should be aware that these chemicals are used in the manufacture of amide-imides.

POLYBUTADIENE

Butadiene is the monomer that is used in the synthesis of polybutadiene, a thermosetting material which is used in electrical applications. Butadiene is a flammable gas which is unsaturated and therefore, polymerizes when subjected to fire conditions. A rapid polymerization can cause violent rupture of the container in which it is housed. Tanks of butadiene must be kept cool during fire conditions.

POLYCARBONATES

Polycarbonates are members of a family of thermoplastic resins relatively new to the plastic industry. These resins are noted for their clarity and their impact resistance. This is the reason they are used as a replacement for window glass. They are also used in appliances, electronics, communications, lighting, and automotive applications.

One of the monomers which can be used in the synthesis of polycarbonates is phosgene, a highly poisonous gas. Fire fighters **should** be aware of the location of phosgene containers when used in the synthesis of polycarbonates. The combustion of polycarbonate does **not** release phosgene as a by-product. Polycarbonates are combustible but can be found in a flame retardant form.

POLYETHYLENE

Polyethylene is a widely used thermoplastic that drips when burning and smells like a burning candle.

Polyethylene is synthesized from ethylene, a colorless, highly flammable and reactive gas. Its flammability range is 3-32 percent. Abrupt heating causes polymerization and rupture of the container. Ethylene is also an anesthetic which causes unconsciousness if inhaled in large amounts.

Polyethylene is found in low, medium, and high density forms. Housewares, toys, containers, garbage bags, wire and cable coating, natural gas piping, and containers for such liquids as soda and milk are only a few of the many uses for polyethylene.

Flame retardant grades of polyethylene are also manufactured. Thermosetting polyethylene can be found in certain applications.

POLYIMIDE

Polyimides are a class of polymers which offer high temperature stability. They are similar to amide-imide in hazardous qualities. Their main use is in fibers, films, and enamels. They can be either thermoplastic or thermosetting.

POLYPROPYLENE

Similar to polyethylene in popularity and performance, polypropylene is a thermoplastic which is found in molding, extrusion, film, and fibers.

It can be made, or reacted, in a variety of ways to form a variety of products. It is synthesized from propylene, a highly flammable gas with properties similar to ethylene. It burns, although not as fast as polyethylene. Flame retardant forms are available.

POLYSTYRENE

Polystyrene is a thermoplastic which is synthesized from the monomer styrene.

Styrene is an aromatic organic liquid until it is polymerized, at which time it becomes a long chain solid. Styrene type monomers can also be mixed with butadiene and acrylonitrite to form ABS plastic.

Polystyrene is popular both as a rigid plastic and as a foam. The fire fighter is probably familiar with it as a low density foam for egg cartons, meat trays, and insulated cups. It is also widely used in toys and packaging, as well as for molded furniture such as TV cabinets. It is combustible; many of the grades sold have flame retardants added to them, however.

SILICONES

Silicones are high temperature polymers which have silicone instead of carbon in the backbone. They are high temperature thermoplastic or thermoset resins, either liquids or solids. They are inherently nonflammable. They are used not only as fluids, but also as silicone rubber and in paints.

POLYVINYLCHLORIDE

Vinyl polymers are prepared from vinyl chloride, which is a flammable gas, or from vinylidene chloride, which is a colorless, flammable liquid. Both are toxic and both polymerize with heat as found under fire conditions. (A Houston incident is described in unit 4.)

The presence of chlorine in the backbone of the polymer makes it inherently flame retardant; however, decomposition of these thermoplastics evolve hydrochloric acid. PVC is used in an extremely wide variety of applications and is second in use only to polyethylene.

POLYURETHANE

The reaction of polyesters with a monomer called an isocyante produces polyure-

thanes. Isocyantes are monomers which are very toxic. Toluene –2, 4-diisocyanate used in the preparation of the majority of polyurethanes is a liquid that is highly toxic. Fire fighters should be aware of the dangers of this moisture reactive chemical.

Urethanes are used both as foams and as solid polymers. Their use as foams for insulation has been widely accepted practice in the past decade. Other applications include flexible elastomers, automotive bumpers, and furniture. They have, however, earned a reputation for producing thick black smoke, with rapid flame propagation, when ignited. Recent advances have produced flame resistant polyurethanes, however.

HAZARDS COMMON TO PLASTICS

Inspection of the data presented for the various plastics listed in tables 20-1 and 20-2 shows similarity with regard to toxic combustion products.

- Plastics containing nitrogen produce ammonia and hydrogen cyanide.

- Plastics containing chlorine produce hydrogen chloride and phosgene. (The precise chemical name for phosgene is carbonyl chloride.)

- Plastics containing bromine produce hydrogen bromide and carbonyl bromide ($COBr_2$).

- Plastics containing fluorine produce hydrogen fluoride and carbonyl fluoride (COF_2).

An incident of plastics presenting toxic hazards to fire fighters occurred January 6, 1970, in Washington, D.C. Fire fighters were called to the scene of a fire in an office where a plastic office machine was burning. The area was filled with smoke and fumes. The decomposition of the plastic, polyvinylchloride, resulted in the formation of hydrogen chloride. The responding fire fighters who were exposed to these fumes became sick. One became severely ill, developed severe pulmonary hemorrhage and edema, and died.

A number of other factors of importance in plastics technology have an indirect relationship on plastic properties. Since they are of secondary importance, relative to the degree of cross-linking and toxic combustion products, they are listed and briefly discussed.

- Plastics often contain fillers to increase shock resistance, reduce shrinkage during cure, provide greater thermal resistance, or simply to reduce the cost of a plastic object. Common fillers are sawdust, asbestos, mica, glass, graphite, and calcium carbonate. In general, the fillers tend to be nonhazardous materials and reduce the hazardous properties according to the proportion present.

- Plasticizers are often added to plastics to increase the stretching and flexibility properties. Most of the industrial plasticizers are esters (formed from acids and alcohols). Commonly used are phosphoric acid and organic acids, including phthalic, stearic, oleic, and adipic. The most commonly used alcohols are glycerol and various glycols.

- Hundreds of different plasticizers are in common use. All are Class B flammable liquids. Most are identified by trade name rather than chemical name. They are usually present in small enough quantities to be of secondary importance in determining hazardous properties.

- Catalysts are used to control the rate of the polymerization reaction. Typical examples are peroxides, sulfuric acid, metals, and metal salts. All are discussed under appropriate headings in this text.

- Small amounts of various dyes, pigments, hardeners, preservatives, and other special purpose additives are usually present in small quantities.

- As with all flammable hazardous materials, surface area is of primary importance in determining the burning rate. Plastic foams or finely divided powders are major flammable hazards. In principle, foams can be made of many plastics by several methods. (1) Mechanical foaming in which a mixture of liquid monomers is whipped to a foam and polymerized to trap the foam structure. (2) Chemical foaming in which a gas evolved in the polymerization reaction is trapped to create a foam structure. (3) Physical foaming in which a foam is formed by adding a volatile solvent to the polymerization mixture. As the low boiling liquid vaporizes, a foam structure is created.

- All foams are flammable hazards by virtue of high surface area. Some are doubly hazardous because of toxic combustion products. Common examples of foam plastics are polyurethanes, with toxic combustion products hydrogen cyanide (HCN) and ammonia (NH_3), polystyrene, cellulose acetate, and foam rubber, with the toxic combustion product sulfur dioxide (SO_2).

- Hazardous situations can occur during the manufacture and processing of plastics. (1) All polymerization reactions are exothermic. The potential of an uncontrollable run-away reaction must be considered in accessing safety. (2) The extent of the flammability hazard depends on the flash point of the plastic monomers. Flash point data for monomers were listed in tables 20-1 and 20-2. (3) Occasionally, some plastic monomers and catalysts have special reactivity and toxicity hazards.

REVIEW

1. A characteristic property of thermoplastics is

   a. Melting.   c. Freezing.
   b. Boiling.   d. Thermal decomposition.

2. The flammable hazardous properties of a thermoplastic are most like a

   a. Gas.   c. Solid.
   b. Liquid.   d. Wood.

3. A characteristic property of thermosetting plastic is

   a. Melting.   c. Burning like a liquid.
   b. Exploding.   d. Burning like wood.

4. An example of a thermoplastic material is

   a. Phenol-formaldehyde resins.   c. Polypropylene.
   b. Wood.   d. Propylene oxide resins.

5. An example of a thermosetting material is

   a. Nylon.   c. Polystyrene.
   b. Teflon.   d. Silicone.

6. Which of the following plastics has the greatest flammable hazard?

   a. Wood.   c. Cellulose acetate.
   b. Coal.   d. Methane.

7. Which of the following plastics forms a toxic combustion product such as hydrogen cyanide?

   a. Thermoplastics.   c. Polyethylene.
   b. Polyamines.   d. Teflon.

8. Which of the following toxic combustion products may be formed by fluorine-containing plastics?

   a. Phosgene.   c. Nitric oxide.
   b. Sodium fluoride.   d. Carbonyl fluoride.

9. Plasticizers are added to plastics to

   a. Reduce the flammable hazard level.
   b. Increase the elastic properties.
   c. Provide a market for trade name products.
   d. Make them burn better.

10. Which of the following plastics are flammable hazardous materials?

    a. Polyurethane foam.   c. Cellulose acetate.
    b. Nitrocellulose.   d. All of these.

11. Plastic foam is hazardous because

    a. It forms HCN and NH_3 which are toxic combustion products.
    b. It has a high surface area.
    c. It is made from flammable monomers.
    d. All of the above.

12. Polycarbonate resins are more hazardous than wood because they

    a. Are thermoplastic materials.   c. Form polymers when heated.
    b. Form CO_2 when burned.   d. None of these.

13. The extent of the flammable hazard of a plastic largely depends on

    a. Unknown factors.   c. Flash point.
    b. The type of plasticizer used.   d. Whether it contains fluorine.

EXPLOSIVE MATERIALS

Unit 21 Principles of Explosives

All explosive processes share the common properties of rapidly producing heat and gases which exert sudden pressure on the surroundings. The greater the heat and gas produced per unit volume, the more powerful the explosive. As discussed earlier in units 7 and 8, explosives are classified as deflagrating explosives or detonating explosives. Detonating explosives are subclassified into (1) high explosives which can be burned without exploding and require a strong mechanical shock or initiator explosion to explode and (2) primary explosives such as those used in initiators or detonators which explode instantly upon ignition.

THERMAL SENSITIVITY

All explosives decompose chemically when they are heated to sufficiently high temperatures. Rapid heating of exothermic materials to temperatures above their thermal decomposition temperature is the condition which produces an explosion.

- An explosive material which decomposes rapidly at 212^6F is considered to be thermally unstable and unsafe.

- An explosive material which can be safely heated to $424°$F is considered to be thermally stable and safe.

- The physical form of the explosive is not crucial. The explosive can be a gas, liquid, or solid. However, denser explosives contain more energy per unit volume and are relatively more explosive for that reason.

Initiating explosives such as lead azide and mercury fulminate do not thermally decompose significantly until a temperature is reached at which they explode suddenly and violently. In contrast, other explosives such as trinitrotoluene (TNT) and nitrocellulose decompose gradually under the same heating conditions and ultimately ignite without explosion.

Another criterion for testing the sensitivity of explosives is by determination of the velocity at which a detonation wave is propagated. Typical values are shown.

decreasing sensitivity		
nitroglycerin	8500	meters/sec.
picric acid	7500	"
trinitrotoluene	7400	"
lead azide	5000	"
mercury fulminate	4800	"
ammonium nitrate	1000	"

The so-called high explosives such as nitroglycerin, picric acid, and trinitrotoluene (TNT) have a high tendency toward detonation and for that reason are limited to military applications. The azides and fulminates are significantly less prone toward detonation, which permits their use in detonators and priming compositions. Ammonium nitrate is the least sensitive of the compounds listed and is often a primary ingredient in ordinary explosives.

Explosives can be ignited by an electric spark. Lead azide and mercury fulminate can be ignited by a weak static electricity spark. Stable explosives such as TNT require an intense electric spark. A fire or open flame is sufficient to instantly detonate an initiating explosive such as lead azide and mercury azide. TNT burns rather than explodes if it is unconfined and present in small quantity.

Shock Sensitivity

All explosives can be exploded by mechanical shock. This is usually referred to as *shock sensitivity*. Initiating explosives such as lead azide and mercury fulminate are detonated by mild shock such as a tap of a pencil. They are extremely hazardous because any movement is apt to detonate them. In contrast, stable explosives such as TNT require a sledge hammer blow to set them off. TNT is a widely used explosive because it can be handled safely. Typical values for shock sensitivity of explosives are listed.

decreasing sensitivity		
mercury fulminate	0.5 inch/pounds
nitroglycerin	1 inch/pounds
picric acid	3.5 inch/pounds
trinitrotoluene	4 inch/pounds
ammonium nitrate	6 inch/pounds

Mercury fulminate and nitroglycerin are the most shock sensitive explosives in ordinary use. Other explosive ingredients such as trinitrotoluene and ammonium nitrate are relatively shock insensitive by comparison.

DETONATING DEVICES

Another aspect of explosives which warrants discussion is the nature and properties of detonating devices. Two basic types of devices are used to set off the main explosive charge without injury to the operating personnel: (1) igniters (squibs, fuses, and delay igniters) which transport a flame to the main explosive charge and (2) detonators (blasting caps and detonating fuses) which transmit a detonation wave to the main explosive charge.

Squibs are small diameter tubes filled with a powder having the desired burning rate properties. Squibs designed to be ignited by a flame contain a slow burning head to allow the operator to reach safety before the flame is transmitted to the more rapidly burning material in the tube. Electric squibs are similar in construction except that ignition is accomplished by means of an electric arc. Electrical squibs are considerably safer because they are more reliable and permit the operator to initiate the squib from a greater distance. Fuses are quite different than squibs in construction. A fuse consists of a black powder core woven into a rope configuration. It differs from a squib in that the fuse is slow burning. A detonation cap at the end or a fast burning powder ignites the main charge. Delay igniters are usually combinations of an electrical igniter and a fuse.

Common blasting caps or detonators are metal cylinders filled with an explosive which is easily detonated. The detonation wave generated by the blasting cap is a reliable means of initiating a primary explosive charge. Electrical blasting caps are identical in construction but are initiated by an electrical arc generated within the cylinder.

REVIEW

1. Which of the following is the most hazardous explosive?

   a. A high explosive.
   b. A primary explosive.
   c. A deflagrating explosive.
   d. A detonating explosive.

2. When an explosion occurs

    a. The explosive is decomposed chemically.
    b. Sudden pressure is exerted on the surroundings.
    c. Heat and gas are produced very rapidly.
    d. All of the above.

3. Which of the following tends to be the most powerful explosive?

    a. A gas.   c. A solid.
    b. A liquid.   d. All are the same.

4. Which of the following explosives is most sensitive as judged by the detonation wave velocity?

    a. Nitroglycerin.   c. Ammonium nitrate.
    b. Mercury fulminate. d. TNT.

5. An explosive which can be detonated by a light blow is

    a. Trinitrotoluene.   c. Picric acid.
    b. Ammonium nitrate.   d. Lead azide.

6. An explosive squib is a type of

    a. Igniting device.   c. Detonating device.
    b. Blasting cap.   d. None of these.

7. Thermal decomposition temperature means the temperature at which a material

    a. Detonates.   c. Chemically breaks down.
    b. Deflagrates.   d. Oxidizes.

8. An explosive ingredient likely to be contained in a military high explosive is

    a. TNT.   c. Both a and b.
    b. Ammonium nitrate.   d. Neither a nor b.

9. Shock sensitivity is most closely related to

    a. Thermal shock.   c. Friction sensitivity.
    b. Pressure sensitivity. d. Mechanical shock.

Unit 22 Nitro Explosives

The nitro explosives manufactured in commercial quantities are based on the availability of cheap, easily obtained starting materials. The basic compounds from which all nitro explosives are made are the hydrocarbons produced by the petroleum industry.

Vapor phase nitration of the saturated hydrocarbons yields nitro explosives of the nitromethane type.

$$CH_4 + HNO_3 \rightarrow CH_3NO_2 + H_2O$$
$$\text{nitromethane}$$

Continued reaction under more severe conditions yields more highly nitrated products.

$$CH_3NO_2 + HNO_3 \rightarrow CH_2(NO_2)_2 + H_2O$$
$$\text{dinitromethane}$$

$$CH_2(NO_2)_2 + HNO_3 \rightarrow CH(NO_2)_3 + H_2O$$
$$\text{trinitromethane}$$

$$CH(NO_2)_3 + HNO_3 \rightarrow C(NO_2)_4 + H_2O$$
$$\text{tetranitromethane}$$

Some of the partially nitrated materials such as nitromethane (CH_3NO_2) are relatively stable materials which are extensively used as solvents, fuels, and synthetic intermediates in industry. The more highly nitrated materials are potential explosives.

The reaction of nitric acid with aromatic hydrocarbons such as benzene, toluene, and phenol produces a similar series of nitro compounds: nitrobenzene, trinitrobenzene, trinitrotoluene (TNT), and trinitrophenol (picric acid). The partially nitrated materials, such as nitrobenzene, are widely used as solvents and chemical intermediates. For example, nitrobenzene is often used as a solvent in lacquers. The more highly nitrated materials are the explosives such as TNT.

The explosive properties of the common nitro explosives are tabulated in table 22-1. The energy content or potential explosive power within a similar series of compounds such as the nitro explosives can be estimated by comparing the percentage of the nitro group present in the molecules. Comparative explosive potential can be estimated as follows for various compounds in a series. The atomic

Compound	Elemental Formula	Calculation of %NO$_2$
trinitrotoluene	$C_7H_5N_3O_6$	$\%NO_2 = \dfrac{3(NO_2) \times 100}{C_7H_5N_3O_6} = \dfrac{3(14+32) \times 100}{7(12) + 5(1) + 3(14) + 6(16)}$ $= \dfrac{138}{227} \times 100 = 60.8$
dinitrotoluene	$C_7H_6N_2O_4$	$\%NO_2 = \dfrac{2(NO_2) \times 100}{C_7H_6N_2O_4} = \dfrac{2(14+32) \times 100}{7(12) + 6(1) + 2(14) + 4(16)}$ $= \dfrac{92}{182} \times 100 = 50.6$
trinitrochlorobenzene	$C_6H_2N_3O_6Cl$	$\%NO_2 = \dfrac{3(NO_2) \times 100}{C_6H_2N_3O_6Cl} = \dfrac{3(14+32) \times 100}{6(12) + 2(1) + 3(14) + 6(16) + 35.5}$ $= \dfrac{138}{247.5} \times 100 = 56.3$

Name	Explosive Properties		
	Energy Content, % NO_2	Brissance	Sensitivity
METHANE DERIVATIVES			
nitromethane	(75.2)	moderate	low
dinitromethane	(89.2)	high	high
trinitromethane	(91.6)	high	high
tetranitromethane	(93.0)	high	high
ETHANE DERIVATIVES			
nitroethane	(61.5)	moderate	low
BENZENE DERIVATIVES			
nitrobenzene	(37.4)	nil	very low
dinitrobenzene	(54.6)	low	low
trinitrobenzene (TNB)	(65.2)	moderate	moderate
TOLUENE DERIVATIVES			
dinitrotoluene	(50.6)	low	low
trinitrotoluene (TNT)	(60.8)	moderate	moderate
PHENOL DERIVATIVES			
trinitrophenol (picric acid)	(60.6)	high	high
RESORCINOL DERIVATIVES			
trinitroresorcinol (styphnic acid)	(56.8)	high	high
ANILINE DERIVATIVES			
trinitroaniline (picramide)	(61.0)	high	high
CHLOROBENZENE DERIVATIVES			
dinitrochlorobenzene	(45.5)	low	low
trinitrochlorobenzene	(56.3)	high	high
BENZOIC ACID DERIVATIVES			
trinitrobenzoic acid	(54.1)	high	high
CRESOL DERIVATIVES			
trinitrocresol (cresolite)	(57.2)	high	high
NAPHTHALENE DERIVATIVES			
tetranitronaphthalene	(60.0)	low	low

Table 22-1. Properties of Nitro Explosives.

weight of each element in the formula must be taken into consideration: N = 14, H = 1, C = 12, O = 16, and Cl = 35.5. (Atomic weight can be found on most copies of the Periodic Table.) This calculation involving only percentages is sufficient to verify that dinitrotoluene is a less powerful explosive than TNT. It also shows that trinitrochlorobenzene is almost as powerful an explosive as is TNT. This technique may be used in estimating the relative explosive power of materials for which handbook data is not available. Data are shown for the common nitro explosives in table 22-1.

Explosive potential is only one factor in determining the hazardous properties of explosive materials. In many respects, thermal and shock sensitivity together with brissance (the tendency to detonate) are even more important. All nitro compounds have sufficiently high energy contents to produce disastrous explosions. They are even more hazardous if the explosion has a greater probability of being initiated by poorly defined factors such as friction, mechanical shock, and slightly elevated temperatures.

The explosives which are used in industry tend to be those which can be handled safely. Maximum explosive power is seldom the primary consideration. For example, hexanitroethane is a much more powerful explosive than TNT, but it is entirely too hazardous to use. The same explosive power can be obtained by using twice as much TNT — which can be handled with a much greater degree of safety.

REVIEW

1. Nitromethane is most widely used as

   a. A fertilizer.
   b. An explosive.
   c. A solvent and racing fuel.
   d. An ingredient in dynamite.

2. Nitrobenzene is most widely used as

   a. A fertilizer.
   b. An explosive.
   c. A racing fuel.
   d. A lacquer solvent.

3. The explosive properties of any nitro compound can be estimated from

   a. $\% NO_2$.
   b. $\% NO_3$.
   c. $\% NO$.
   d. $\% O$.

4. The comparative explosive potential in a hypothetical compound having the elemental formula $C_9 H_7 (NO_2)_2$ is

   a. 19.6.
   b. 22.2.
   c. 44.4.
   d. 88.8.

5. TNT is a more widely used explosive than hexanitroethane because

   a. TNT is more brissant.
   b. TNT is less brissant.
   c. TNT is more thermally sensitive.
   d. TNT is more shock sensitive.

6. An example of thermal shock is

   a. A rapid increase in temperature.
   b. Friction.
   c. An explosion.
   d. All of these.

7. An example of shock sensitivity is

   a. An unexpected explosion.
   b. An explosion initiated by a hammer blow.
   c. A sonic boom.
   d. A loud noise.

Unit 23 Nitric Ester Explosives

The nitric ester explosives are formed by the reaction of nitric acid with an alcohol.

$$CH_3O \; \textcircled{H} \; + \; \textcircled{H} \, N \, \textcircled{O}_3 \; \rightarrow \; CH_3ONO_2 + H_2O$$

The nitro explosives discussed in unit 22 are formed from the reaction of nitric acid with a hydrocarbon.

$$C \, \textcircled{H}_4 \; + \; \textcircled{H} \, N \, \textcircled{O}_3 \; \rightarrow \; CH_3NO_2 + H_2O$$

The important difference in the nitric ester explosives is in their higher oxygen content.

The common nitric ester explosives and their explosive properties are listed in table 23-1. The student probably wonders why TNT is safer than nitroglycerin and why nitrocellulose is safer than nitroglycerin. Knowledge of the difference in energy content of a nitric ester (ONO_2) explosive compared to a nitro (NO_2) explosive removes much of the mystery and confusion about these explosives.

The explosive power or energy content of nitro and nitric ester explosives can be directly compared from the data presented in tables 23-1 and 22-1. To aid in this comparison, table 23-1 lists the energy content both as $\%NO_3$ and equivalent $\%NO_2$. The equivalent $\%NO_2$ numbers are obtained in this way:

$$\text{equivalent } \%NO_2 = \%NO_3 \times \frac{NO_3}{NO_2} =$$

$$82\% \times \frac{62}{46} = 82\% \times 1.35 = 111\%$$

It can be seen that the nitric ester explosives have a much higher explosive energy

Name	Structure	$\%NO_3$	Equivalent $\%NO_2$	Brissance	Sensitivity
methyl nitrate	CH_3ONO_2	80.5	108	high	high
ethyl nitrate	$CH_3CH_2ONO_2$	68.0	91.5	high	high
ethyleneglycol dinitrate	$\begin{matrix} CH_2 - CH_2 \\ \mid \qquad \mid \\ ONO_2 \quad ONO_2 \end{matrix}$	81.5	110	high	high
nitroglycerin	$\begin{matrix} CH_2ONO_2 \\ \mid \\ CH \; ONO_2 \\ \mid \\ CH_2ONO_2 \end{matrix}$	82.0	111	high	high
nitrocellulose	variable composition but approximately $\left[\begin{matrix} CH_2 \\ \mid \\ HCONO_2 \\ \mid \\ HCONO_2 \quad O \\ \mid \\ HCONO_2 \\ \mid \\ CH \\ \mid \\ HCO - \end{matrix} \right]_X$	63.3	85.2	high	moderate
ammonium nitrate	NH_4NO_3	73.0	98.5	high	high

Table 23-1. Structure and Explosive Properties of Nitric Ester Explosives.

content than the nitro explosives. This ratio is often expressed as the TNT equivalent of an explosive. Some of the TNT equivalents of the nitric ester explosives are listed.

- nitroglycerin $= \dfrac{111}{60.8} = 1.83$

- ethyleneglycol dinitrate $= \dfrac{110}{60.8} = 1.82$

- ammonium nitrate $= \dfrac{98.5}{60.8} = 1.62$

- nitrocellulose $= \dfrac{85.2}{60.8} = 1.40$

All have a significantly higher energy content than TNT. TNT is a convenient reference point since it is a relatively safe explosive. All explosives are hazardous, but TNT is relatively safe compared to the nitric ester explosives.

Among the nitric ester explosives, nitrocellulose has the lowest energy content and is the least explosive. It bridges the gap between TNT and nitroglycerin.

Explosives are made from fully nitrated cellulose. Lacquers and films are made from dinitrated cellulose. Plastics are made from mixtures of the mono- and dinitrated cellulose.

To many students, ammonium nitrate may appear to be out of place classified as a nitric ester. It is formed by the reaction of ammonium hydroxide with nitric acid. Viewed

$$NH_4O\ \textcircled{H}\ +\ \textcircled{H}\ N\ \textcircled{O}_3\ \rightarrow\ NH_4ONO_2\ +\ H_2O$$

in terms of chemical similarity, NH_4OH and CH_3OH are very much alike. The formula for ammonium nitrate is ordinarily written as NH_4NO_3 rather than NH_4ONO_2. The same is true of all of the nitric esters. Methyl nitrate is usually written as CH_3NO_3 rather than CH_3ONO_2. The ONO_2 form is more chemically correct, however.

Ammonium nitrate receives much attention as an explosive hazard because it is used in huge quantities in fertilizers and basic chemical industries. It is transported in shipload and trainload quantities. Much can be said about its hazardous properties but all can be summarized in two short statements.

- It has a TNT equivalent of 1.62.

- It behaves like any other nitric ester, such as nitroglycerin.

REVIEW

1. An example of a nitric ester explosive is

    a. Ammonium nitrate.
    b. Nitromethane.
    c. Nitroglycerin.
    d. All of the above.

2. The TNT equivalent of nitroglycerin is

    a. Much higher than ethyleneglycol dinitrate.
    b. Much lower than ethyleneglycol dinitrate.
    c. About the same as ethyleneglycol dinitrate.
    d. Cannot be compared to ethyleneglycol dinitrate.

3. Ammonium nitrate is used in explosives because

    a. It has a high TNT equivalent.
    b. Surplus ammonium nitrate is available from the agriculture business.
    c. It is a relatively safe explosive because it is water soluble.
    d. It is nontoxic to humans.

4. The proper empirical formula for a nitric ester group is

    a. N_2O_4.   c. NO_3.

    b. NO_2.   d. NH_4.

5. Lacquers based on nitrocellulose formulations are primarily

    a. Explosive hazards.   c. Toxic hazards.

    b. Flammable hazards.   d. Nonhazardous.

6. Nitric esters are made by the reaction of

    a. Alcohols and nitric acid.   c. Nitrates and nitric acid.

    b. Amines and nitric acid.   d. All of these.

7. Which of the following nitric esters are most similar?

    a. Methyl nitrate and ammonium nitrate.

    b. Methyl nitrate and nitroglycerin.

    c. Nitrocellulose and nitroglycerin.

    d. Methyl nitrate and ethyl nitrate.

Unit 24 Other Specialty Explosives

In this unit a wide variety of specialty explosives will be considered for which there is little chemical similarity and consequently generalizations of the type applied to the nitro and nitric ester series are not possible.

DYNAMITE

Dynamite is an explosive mixture used in controlling oil well fires, in construction, in mining, and in agriculture. It is usually detonated with an electrical blasting cap. Unlike the explosive materials discussed thus far, dynamite is a mixture of explosives and inert additives. The properties characteristic of dynamite are conferred to a large extent by the inert component rather than the explosive which may be nitroglycerin, ammonium nitrate, or sodium nitrate. Some typical dynamite compositions are discussed.

Straight Dynamite

Straight dynamite contains nitroglycerin as the primary explosive. A typical approximate formulation is shown.

> 40% nitroglycerin
> 45% sodium nitrate
> 14% wood pulp, flour or similar material
> 1% antiacids such as $CaCO_3$

Straight dynamite has been largely replaced by compositions containing smaller amounts of nitroglycerin. The basis for this replacement stems from the fact that large volumes of toxic gases are emitted upon detonation of straight dynamite. This hazard restricts its use to underwater blasting operations.

Ammonia Dynamite

Ammonia dynamite contains ammonium nitrate as the major explosive ingredient. A typical approximate formulation is shown.

> 30% ammonium nitrate
> 40% sodium nitrate
> 15% nitroglycerin
> 10% wood pulp, flour, or similar material
> 4% sensitizer
> 1% antiacid

Incorporation of ammonium nitrate as the primary explosive reduces the sensitivity so much that a standard blasting cap will no longer detonate the mixture. A sensitizer is added to the formulation to regain the lost detonation capability. The most commonly used sensitizer is sulfur; others which are used occasionally are diphenylamine, dinitrotoluene, nitronaphthalene, and aluminum powder. Ammonium nitrate is a very hygroscopic material (absorbs moisture from atmosphere) and appropriate measures must be taken in the dynamite formulation to counteract this problem. A hygroscopic dynamite formulation is unsuitable for use in wet or humid areas.

Ammonia dynamites are always manufactured in the form of coarse grains which have little tendency to cement together when moisture is absorbed. In contrast, a fine-grained formulation would be cemented together under the same conditions. Moisture absorption in a coarse-grain formulation does not lead to a change in explosive properties because the burning surface remains unchanged.

Ammonia dynamites are manufactured in 15 to 60 percent grades. The percentage indicates the amount of ammonium nitrate in a given formulation.

Gelatin Dynamite

Gelatin dynamite contians a gelatinous mixture made by dissolving nitrocellulose in nitroglycerin. This is a dilute solution of nitrocellulose in nitroglycerin known as *collodion,* a common material in chemical laboratories and hazardous in its own right. Gelatin dynamite does not actually contain gelatin; the so-called gelatin is a descriptive term for the gel formed by the solution of nitrocellulose in nitroglycerin. A typical approximate formulation is shown.

40% sodium nitrate

50% gelatin

9% wood pulp, flour, or similar material

1% antiacid

Gelatin dynamites have a number of major advantages over other types and are manufactured in grades ranging from 20%-90% gelatin. Other variants of the basic gelatin dynamite formulation are (1) blasting gelatin which is 100% nitroglycerin-nitrocellulose gel (It does not contain the other typical dynamite ingredients such as ammonium nitrate, sodium nitrate, and wood pulp. It is a translucent, rubbery material which is widely used in submarine applications because of its excellent moisture resistant properties.), (2) semigelatin which contains only a small amount of gelatin sufficient to improve the moisture absorption properties of ammonia dynamites (this is the dynamite most commonly used in rock blasting operations), and (3) ammonia gelatin which is an ammonia dynamite in which the nitroglycerin component is replaced by gelatin. (The advantages conferred to the ammonia dynamite are similar to those of any gelatin dynamite. This dynamite is manufactured in 30%-90% grades.)

BLACK POWDER

Black powder is a mixture of potassium nitrate, sulfur, and charcoal which was widely used in the past. Currently, it has very limited use — mainly by sportsmen interested in muzzle-loading guns. A typical approximate composition is shown.

75% potassium nitrate

12.5% sulfur

12.5% charcoal

Black powder was a reasonably good explosive, with some desirable properties: low detonation rate, low shock sensitivity, reasonably energetic, and made from inexpensive materials. One major factor made it a highly dangerous and unreliable explosive — the hygroscopicity of potassium nitrate and charcoal. Moisture absorption caused the black powder to cake and the ignition characteristics to alter in an unpredictable manner. The result was unpredictable ignition and misfires. Black powder is seldom used for this reason. Historically, it was the first of the modern explosives. Modern dynamite evolved from black powder.

PYROTECHNICS

Pyrotechnics are largely based on black powder compositions to which additional materials are added to achieve special effects. The hazards associated with pyrotechnics are similar to those of black powder. Many of the additives in pyrotechnics contribute toxic combustion products hazards as well as increase the unreliability of ignition. Some typical additives used to obtain special effects in fireworks are shown.

- Red color by addition of strontium salts

- Yellow color by addition of socium salts

- Blue color by addition of copper-mercury salt mixtures

- Green color by addition of barium salts
- White color by addition of antimony or arsenic salts
- Brilliant white color by addition of aluminum and magnesium powder
- Sparks by addition of barium and lead nitrates

AZIDES AND FULMINATES

Azides and fulminates are the explosives most commonly used in initiators and primers. All of the fulminates and azides are highly shock, thermal, and friction sensitive. Although these properties place them among the most hazardous of explosive materials, these are the very properties which ensure reliable functioning of igniting and priming devices.

Mercury Fulminate

The only fulminate compound which is used in explosive applications is mercury fulminate, $Hg(OCN)_2$. The term fulminate is jargon used in the explosive industry. The correct chemical name is mercury cyanate. For example, the related compound $NaOCN$ would be located in handbooks under the name sodium cyanate rather than sodium fulminate.

Mercury fulminate is widely used in blasting caps. In the blasting cap application, it is sensitive to moisture and is less easily detonated. This is a major disadvantage in blasting caps. On the other hand, this property is used to advantage in shipping mercury fulminate. Water functions as a desensitizer in this case.

Silver Fulminate

Another fulminate which is also likely to be encountered is silver fulminate. It is formed in all mirror-silvering baths.

Although many mirror-silvering formulations are possible, most utilize a silver salt such as silver nitrate, a reducing agent such as sugar or tartaric acid, and ammonium hydroxide. On standing, silver fulminate forms in these solutions and for that reason, the solutions must be disposed of promptly. Silver fulminate is just as hazardous as mercury fulminate.

Azides can be used interchangeably with fulminates in initiator and primer applications. Lead azide (PbN_3) and mercury azide (HgN_3) are the ones commonly used. Azides are slightly more sensitive than fulminates but the overall hazardous properties are similar.

EXPLOSIVE DETONATORS

In detonators, explosives are used between the initiator or primacord assembly to ensure reliable ignition of the main explosive charge. The highly sensitive materials such as the fulminates and azides are impractical to use in the larger quantity necessary in the detonator. A less sensitive material which is still capable of detonating reliably is necessary for this purpose. They tend to lie between the sensitivity range of mercury fulminate and nitroglycerin. Some of the explosives commonly used as detonators are listed.

- Cyclotrimethylenenitramine (also called cyclonite, RDX)
- Trinitrophenylmethylnitramine (also called tetryl)
- Trinitrophenylnitramine
- Lead trinitroresorcinate (also called lead styphnate)

Another specialized type of detonating material is primacord or detonating cord. It is filled with pentaerythritoltetranitrate (PETN). Within the primacord, the PETN propagates a detonation with extreme reliability and at a velocity of 300 feet per second. The hazardous properties are very similar to those of nitroglycerin or any other nitric ester.

REVIEW

1. The primary difference between a straight dynamite and an ammonia dynamite is

    a. The higher nitroglycerin content of ammonia dynamite.
    b. The higher sodium nitrate content of ammonia dynamite.
    c. The lower nitroglycerin content of ammonia dynamite.
    d. The higher ammonia content of ammonia dynamite.

2. The primary difference between straight dynamite and gelatin dynamite is

    a. The nitrocellulose content.   c. The ammonium nitrate content.
    b. The nitroglycerin content.   d. The sodium nitrate content.

3. Gelatin dynamite has the following property.

    a. It has a low moisture absorption.
    b. It has a high density.
    c. It looks like a synthetic plastic rather than a granular powder.
    d. All of the above.

4. Blasting gelatin has a high content of

    a. Nitroglycerin.   c. Sodium nitrate.
    b. Ammonium nitrate.   d. Sulfur.

5. If the potassium nitrate component of black powder were replaced by strontium nitrate

    a. The black powder would not explode.
    b. The black powder would burn with a blacker flame.
    c. The black powder would burn with a red flame.
    d. The black powder would not be hygroscopic.

6. The hazardous properties of pyrotechnics are most similar to those of

    a. Straight dynamite.   c. RDX.
    b. Gelatin dynamite.   d. Black powder.

7. The azide compound most likely to be encountered in explosive applications is

    a. Silver azide.   c. Lead azide.
    b. Nitroazide.   d. None of these.

8. An explosive hazardous fulminate likely to be encountered is

    a. Silver fulminate.   c. Potassium fulminate.
    b. Lead fulminate.   d. Calcium fulminate.

9. Primacord contains which of the following explosive materials?

    a. Mercury fulminate.   c. PETN.
    b. Lead azide.   d. Lead styphnate.

SECTION 6 REACTIVE MATERIALS

Unit 25 Peroxides

Hydrogen peroxide is the best known of the peroxides because of its extensive use as an antiseptic. It is most commonly encountered as a 3-percent solution in water. It is useful as an antiseptic because as an unstable, reactive material, it produces oxygen during the decomposition process.

$$2H_2O_2 \rightarrow 2H_2O + O_2 + heat$$

All peroxides are inherently unstable materials which decompose exothermically.

- They decompose catalytically in contact with other materials. Traces of metals and metal salts are efficient decomposition catalysts.

- They are strong oxidizing agents and decompose in a wide variety of oxidation-reduction reactions. The primary modes of decomposition of all peroxides are similar.

$$2H_2O_2 \xrightarrow[\text{decomposition}]{\text{catalytic}} 2H_2O + O_2$$
$$C + 2H_2O_2 \xrightarrow{\text{oxidation}} 2H_2O + CO_2$$

Hydrogen peroxide is an oxidizing agent. When it decomposes spontaneously, heat is generated and oxygen is liberated. In the dilute 3-percent solution commonly used in household disinfectants, the heat liberation is a minor factor and the hazard associated with oxidation reactions is low.

Ten- and thirty-percent concentrations of hydrogen peroxide are commercially available at all scientific supply houses and are widely used in laboratory and industrial processes. At the ten-percent concentration, hydrogen peroxide is a hazardous material; near explosive combustion conditions occur when this concentration is poured on grease on a concrete driveway. At a thirty-percent concentration, an explosion would result under the same conditions. Concentrations of hydrogen peroxide to about ninety percent are used in the rocket and other industries. As the percentage concentration of hydrogen peroxide is increased, the hazard associated with the material is increased proportionally.

PEROXIDE REACTIONS

The reactions of peroxides present a number of special problems.

- Reactions of a peroxide with carbon dioxide produces oxygen. In a fire situation, a cyclical or regenerative system may be set up which is difficult to extinguish.

$$2Na_2O_2 + 2CO_2 \rightarrow 2Na_2CO_3 + O_2$$
$$wood + O_2 \rightarrow CO_2 + H_2O$$
$$2CO_2 + 2Na_2O_2 \rightarrow 2Na_2CO_3 + O_2$$

- Metal peroxides are reactive with water. This precludes the use of water in combating metal peroxide fires.

- Metal peroxides are thermally unstable.

- Metal peroxides are brissant explosives and can detonate.

- All peroxides are unstable materials which slowly decompose during storage. Hydrogen peroxide decomposes to yield water and oxygen.

$$H_2O_2 \rightarrow H_2O + O_2$$

Acetyl peroxide, however, decomposes to yield substituted hydrocarbons and oxygen.

acetyl peroxide → acetone + acetaldehyde + oxygen

$$CH_3\underset{O}{\overset{\parallel}{C}}-OO-\underset{O}{\overset{\parallel}{C}}CH_3 \rightarrow CH_3\underset{O}{\overset{\parallel}{C}}CH_3 + CH_3\underset{O}{\overset{\parallel}{C}}H + O_2$$

The particular substituted hydrocarbons which are formed during decomposition of a peroxide are not so important as the fact that they are produced together with oxygen. This is obviously a situation which is a potential flammable and explosive hazard.

STORAGE

For the above stated reason, peroxide containers are vented to prevent the oxygen and organic vapor from accumulating. Storage and transportation demands adequate ventilation to avoid the buildup of flammable or explosive vapors.

USES

Organic peroxides are commonly used as initiators in the polymerization process. The most familiar application is perhaps in household adhesives or boat fiberglass. An initiator, which is often a peroxide, is added to the mixture of monomers and the desired polymer results. The peroxide functions in the polymerization reaction by decomposing to yield a free radical, which in turn reacts with the monomer to form a plastic.

HAZARDOUS PROPERTIES

The hazardous properties of organic peroxides are directly related to the property of easily decomposing to form a free radical.

- Organic peroxides are highly exothermic. Decomposition to yield a free radical liberates heat.

- Organic peroxides are highly reactive. The free radicals formed during decomposition are the reactive species. This reactivity of free radicals is the basis of the use of organic peroxides in the plastics industry.

- Organic peroxides are strong oxidizing agents.

- Organic peroxides are highly toxic because of their chemical reactivity.

KINDS OF ORGANIC PEROXIDES

Two types of organic peroxides are encountered. They can be viewed as derivatives of hydrogen peroxide. When one hydrogen of hydrogen peroxide is replaced, a hydroperoxide is formed. When both hydrogens are replaced an organic peroxide is formed.

- Hydroperoxides —OOH

| CH_3OOH | methyl hydroperoxide |
| 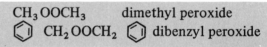 CH_2OOH | benzyl hydroperoxide |

- The so-called   —OO— organic peroxides

| CH_3OOCH_3 | dimethyl peroxide |
| CH_2OOCH_2 | dibenzyl peroxide |

The primary difference in the hazardous properties of the inorganic peroxides (H_2O_2 and metal peroxides), the hydroperoxides, and the organic peroxides is the difference in carbon-hydrogen content.

- H_2O_2 is not flammable. It is an unstable reactive material but combustion does not result unless a hydrocarbon material is present. Recall that oxygen does not burn, but oxygen-gasoline mixtures do.

- The hydro- and organic peroxides burn because they contain the elements C, H, and O. Their flammable and explosive properties are very similar to the nitro explosives. They carry their own oxygen and are therefore potential explosives.

- The properties of the hydro- and organic peroxides can be compared to those of TNT. They are shock and friction sensitive.

The oxygen content of the common peroxides is given in table 25-1. One hundred

Name	Formula	% Oxygen	
		Total	Available
Inorganic Peroxides			
hydrogen peroxide	H_2O_2		
3%		2.8	1.4
10%		9.4	4.7
50%		47	23.5
100%		94	47
lithium peroxide	Li_2O_2	70	35.5
sodium peroxide	Na_2O_2	41	20.5
potassium peroxide	K_2O_2	29	14.5
calcium peroxide	CaO_2	44	22
barium peroxide	BaO_2	19	9.5
Organic Peroxides			
acetyl peroxide	CH_3C—OO—CCH_3 (with O below each C)	25	12.5
benzoyl peroxide	(benzene ring)—C—OO—C—(benzene ring) with O below each C	13	6.5
Organic Hydroperoxides			
cumene hydroperoxide	(benzene ring)—C—OOH with CH_3 above and CH_3 below	21	10.5
tertiary butylperoxide	H_3C, H_3C, H_3C > C—OOH	17	8.5

Table 25-1. Properties of Peroxides.

percent hydrogen peroxide has an oxygen content of 94%. It is important to note that only half of this oxygen is available for oxidation reactions. The remaining half is used in formation of water which under most circumstances is a nonhazardous material

$$2H_2O_2 \rightarrow 2H_2O + O_2$$

FIRE FIGHTING PROCEDURES

Peroxides are transported with the **Dangerous** label on the vehicle along with an **oxidizer** label. Fires or accidents involving peroxides warrant extreme care. Since peroxides supply their own oxygen, flame propagation is extremely rapid and in some cases detonation occurs. Fires involving peroxide should be handled with water. Depending on the amount of peroxides present, the use of fog for cooling or master streams for containment may be desirable. Peroxide fires are extremely difficult to extinguish and the safety of the surrounding area should be a prime concern. Overhaul operations should not start until all areas are cool and free from heat.

REVIEW

1. Household hydrogen peroxide used as an antiseptic is
   a. A 2% solution of H_2O_2 in water.
   b. A 3% solution of H_2O_2 in water.
   c. A 5% solution of H_2O_2 in water.
   d. Not diluted with water.

2. All peroxides are
   a. Thermally unstable.   c. Highly reactive.
   b. Catalytically unstable.   d. All of these.

3. Hydrogen peroxide is
   a. An oxidizing agent.   c. A corrosive.
   b. A reducing agent.   d. An acid.

4. Hydrogen peroxide decomposes by which of the following reactions?
   a. $H_2O_2 \rightarrow H_2 + O_2$.   c. $H_2O_2 \rightarrow O^- + H_2O$.
   b. $H_2O_2 \rightarrow H^+ + H_2O$.   d. $2H_2O_2 \rightarrow O_2 + 2H_2O$.

5. Hydrogen peroxide is chemically most similar to
   a. Na_2O_2.   c. H_2O.
   b. NaOH.   d. H_2.

6. Metal peroxides are
   a. Chemically reactive with water.
   b. Chemically reactive with carbon dioxide.
   c. Chemically reactive with hydrogen.
   d. All of the above.

7. An example of a thermally unstable hazardous material is
   a. Sodium peroxide.   c. Oxygen.
   b. Acetone.   d. Hydrogen.

8. Organic peroxides are more hazardous than hydrogen peroxide because
   a. They decompose to yield water and oxygen.
   b. They decompose to yield organic compounds and oxygen.
   c. They decompose to yield hydrogen and oxygen.
   d. None of the above.

9. Organic peroxides are most likely to be encountered in
   a. Photographic development laboratories.
   b. Rocket launch sites.
   c. Medical offices.
   d. Plastics manufacturing.

10. Peroxides are not reactive with
    a. Hydrogen.   c. Oxygen.
    b. Methane.   d. Carbon.

11. Hydrogen peroxide is
    a. Not flammable.   c. A reducing agent.
    b. Flammable.   d. Nontoxic.

12. Which of the following metal peroxides presents the greatest oxidation hazard?
    a. Sodium peroxide.   c. Calcium peroxide.
    b. Potassium peroxide.   d. Lithium peroxide.

Unit 26 Hydrazines

Only three hydrazine derivatives are likely to be encountered in large quantities. All

hydrazine	monomethylhydrazine (MMH)	unsymmetrical dimethylhydrazine (UDMH)
$\begin{array}{cc} H & H \\ \diagdown & \diagup \\ N\!-\!N \\ \diagup & \diagdown \\ H & H \end{array}$	$\begin{array}{cc} H_3C & H \\ \diagdown & \diagup \\ N\!-\!N \\ \diagup & \diagdown \\ H & H \end{array}$	$\begin{array}{cc} H_3C & H \\ \diagdown & \diagup \\ N\!-\!N \\ \diagup & \diagdown \\ H_3C & H \end{array}$

are amines and as such possess all of the properties of amines previously discussed in unit 19. They are singled out for discussion here not so much because of unusual hazardous properties, (although they are flammable, reactive, corrosive, and toxic) but because of their extensive use in the rocket industry as rocket fuels. For that reason, the probability of encounter is rather high.

Both monomethyl- and dimethylhydrazine have hazardous properties similar to all organic amines. Selected properties together with those of hydrazine are shown in table 26-1.

These particular substituted hydrazines are the only ones manufactured in large quantity. Hydrazine is made entirely from ammonia and sodium hypochlorite. The other two hydrazine derivatives, monomethylhydrazine and uns-dimethylhydrazine, are partially made from ammonia. Since ammonia is the least expensive starting material, the above three hydrazine derivatives are among the compounds of this type most often manufactured and most often encountered.

USE OF HYDRAZINE

Hydrazine itself has a number of special properties. As a monopropellant in rocketry, it is a highly exothermic material which is catalytically decomposed by metals to form nitrogen, hydrogen and ammonia. Hydrazine decomposes to yield a gas at a high temperature and pressure which can be expanded through a rocket nozzle to provide rocket thrust. The unique feature of the hydrazine monopropellant rocket is related to the fact that spontaneous decomposition can be achieved by simply spraying hydrazine on a metal surface. This property permits design of a simple propulsion system in which rocket thrust can be controlled by a single off-on valve.

HAZARDS OF HYDRAZINE

The explosive hazards associated with hydrazine are fundamentally the same as those which make it an effective propellant.

- Spontaneous decomposition induced by heat or contact with metals is an ever present hazard. In a confined volume, an explosion results. In an unconfined volume, a fire results.

	MP, °F	BP, °F	Flash Point, °F	Flammable Limit % in air		Toxicity
				Lower	Upper	
hydrazine	34	236	100	- - -	- - -	high
MMH	- - -	189	61	2.5	100	high
UDMH	- - -	146	about 30	2	100	high

Table 26-1. Properties of the Hydrazines.

Name	Structure	Hazardous Properties	Occurrence
phenylhydrazine	$NHNH_2$  ⬡	Same as any amine	Common laboratory reagent
hydroxylamine	NH_2OH	Same as any amine	Common laboratory reagent
hydrazine hydrate	$N_2H_4 \cdot H_2O$	Same as any amine	Common laboratory reagent
2, 4-dinitrophenylhydrazine	$NHNH_2$  ⬡ NO_2 NO_2	Explosive, contains both reducing and oxidizing groups on the same molecule	Common laboratory reagent

Table 26-2. Other Compounds Related to Hydrazine.

- The decomposition proceeds via a free radical process. The reaction is violent because of the high reactivity of all free radicals. The magnitude of this hazard is similar to that of organic peroxides and TNT.

PROPERTIES OF SIMILAR MATERIALS

The other hydrazines, MMH and UDMH do not catalytically decompose. In general, any hydrazine compound containing carbon atoms cannot be catalytically decomposed. Other materials which are chemically similar to hydrazine are listed in table 26-2. Phenyl-

hydrazine, hydroxylamine, and hydrazine hydrate are common laboratory reagents used for chemical analysis. All are amines and therefore their hazardous properties are those of any amine. Their boiling point and flash point properties are similar to those of monomethyl hydrazine.

FIRE FIGHTING CONSIDERATIONS

Most hydrazines are miscible to an extent with water. Any encounter with spills should be diluted with copious amounts of water. Fires involving hydrazines should be fought with large amounts of water.

REVIEW

1. All hydrazine compounds are similar to what class of chemical compounds with respect to hazardous properties?

   a. Amides.
   b. Nitro compounds.
   c. Nitric esters.
   d. Amines.

2. All hydrazine compounds are

   a. Reducing agents.
   b. Oxidizing agents.
   c. Acids.
   d. Aldehydes.

3. Hydrazine is catalytically decomposed by metals according to the following reaction:

   a. $2N_2H_4 \rightarrow 4NH_3 + H_2$.
   b. $2N_2H_4 \rightarrow N_2 + H_2 + 2NH_3$.
   c. $N_2H_4 \rightarrow N_2 + 2H_2$.
   d. None of these.

4. Hydrazines may be

   a. Reactive.   c. Toxic.
   b. Corrosive.   d. All of these.

5. Both hydrazine and monomethylhydrazine are

   a. Catalytically decomposed.   c. Flammable.
   b. Catalysts.   d. All of these.

Unit 27 Miscellaneous Reactive Materials

A wide variety of materials are hazardous because of their special properties. Classification under a miscellaneous heading does not imply that they are either less hazardous or that they are less likely to be encountered.

Considerable detail has been presented for the major classes of materials. The fundamental chemical background established to this point permits discussion of a wide variety of materials without the necessity of the extensive background discussions required previously.

HYDRIDES

The hydrocarbons and substituted hydrocarbons are actually hydrides of carbon. Their properties are not typical of most hydrides. They are exceptions to the general chemical behavior of hydrides. Presumably, hydrocarbons are prevalent because of their relative stability. In contrast, most hydrides are highly reactive, hazardous materials.

- Hydrides react with water to form hydrogen and bases.

  $$2NaH + 2H_2O \rightarrow 2NaOH + 2H_2$$

- Hydrides react with acids to form metal chlorides and hydrogen.

  $$NaH + HCl \rightarrow NaCl + H_2$$

- Hydrides react with oxidizing agents.

  $$2NaH + O_2 \rightarrow Na_2O + H_2O$$

- Hydrides react with halocarbons.

  $$4NaH + CCl_4 \rightarrow 4NaCl + CH_4$$

- All hydrides are strong reducing agents.

Metal Hydrides

The Group I metal hydrides all react violently with water and spontaneously with moist air. In a practical sense, all are so reac-

LiH, NaH, KH, RbH, CsH

tive that relative reactivity is of little consequence. From the viewpoint of periodic relationships, hydrogen may be considered to be hydrogen hydride HH.

The Group II and Group III hydrides are somewhat less reactive but not significantly so. They are extremely reactive materials by any standard.

BeH_2, MgH_2, CaH_2, SrH_2, BaH_2 AlH_3

Most of the other metals form hydrides. They are generally less reactive but are still clearly in the hazardous reactive material category. Most are not likely to be encountered except in research laboratories.

On the other hand, mixed metal hydrides are extensively used. Some of the more common are shown. They fall into the same

$LiAlH_4$	$(LiH \cdot AlH_3)$	lithium aluminum hydride
KBH_4	$(KH \cdot BH_3)$	potassium borohydride
$NaAlH_4$	$(NaH \cdot AlH_3)$	sodium aluminum hydride
$NaBH_4$	$(NaH \cdot BH_3)$	sodium borohydride

hazardous category as the other metal hydrides. This is to be expected if they are viewed as a mixture of a Group I and a Group III hydride, although the different appearance of the formula might lead to the conclusion that the reactivity is different.

Hazard

Metal hydrides should be kept free from water during any spills or fire. The reaction with water is violent.

Nonmetal Hydrides

The common nonmetal hydrides are shown. As might be anticipated, the proper-

Group V	NH_3	ammonia
	PH_3	phosphine
Group VI	H_2O	water
	H_2S	hydrogen sulfide

ties of the nonmetal hydrides differ considerably from those of the metal hydrides. Without going into chemical details, it may be said that the hazardous properties of the nonmetal hydrides are not uniquely related to chemical reactivity. By comparison to the metal hydrides, the reactivity of the nonmetal hydrides is low.

- They are weak reducing agents.

- They dissolve in water rather than react with the liberation of hydrogen.

Hazard

On the other hand, the nonmetal hydrides as a group tend to be highly toxic gases. Ammonia, phosphine, and hydrogen sulfide gases are highly toxic. This general property of toxicity is also shared by the semimetal hydrides.

Boron hydrides (the boranes)	
B_2H_6	diborane
B_5H_9	pentaborane
$B_{10}H_{14}$	decaborane
SiH_4	silane
AsH_3	arsine
TeH_2	tellurium hydride

Of the semimetal hydrides, the boranes are probably best known because of the interest in their use as high energy rocket propellants. For various reasons, among them hazardous properties and toxicity, the boranes are no longer seriously considered as potential propellant materials. The boranes find some industrial application as corrosion inhibitors. The other semimetal hydrides, SiH_4, AsH_3, and TeH_2, are likely to be encountered only in

research laboratories. The semimetal hydrides are highly hazardous materials.

- They are extremely toxic.

- They are all gaseous except decaborane.

- They are exothermic.

- They are pyrophoric in air.

- They react violently with water, oxidizers, and halocarbons.

CARBIDES

Metals react with carbon when heated to sufficiently high temperature to form binary compounds called carbides. The alkali metals (Group I) and alkaline earth metals (Group II) form carbides corresponding to the general formulas M_2C_2 and MC_2, respectively.

$$2Na + C \rightarrow Na_2C_2$$
$$Ca + 2C \rightarrow CaC_2$$

The alkali and alkaline earth carbides react similarly and form acetylene ($HC \equiv CH$) on reaction with water. This is the basis of the carbide lamps traditionally used by miners. It is also the basis for the hazardous properties of these carbides because acetylene is a highly flammable, exothermic gas. The other reaction product, calcium hydroxide, is a potential corrosive hazard

$$CaC_2 + 2H_2O \rightarrow HC \equiv CH + Ca(OH)_2$$

The alkali and alkaline earth carbides are often called acetylides because they form acetylene by reaction with water.

Beryllium and aluminum form the compounds Be_2C and Al_4C_3, which give methane upon hydrolysis. Again a flammable hazardous gas, methane, is formed.

Nonhazardous Carbides

Most of the other carbides are distinctly nonhazardous. They are characterized by high

Class	Example	Hazard
Sulfides	Na_2S CaS Fe_2S_3 P_4S_3	Ordinarily nonhazardous but yield hydrogen sulfide by reaction with acids ($Na_2S + 2HCl \rightarrow H_2S + 2NaCl$) and sulfur dioxide when burned in air. ($Na_2S + O_2 \rightarrow Na_2O + SO_2$)
Phosphides	Ca_3P_2	Yields phosphine by reaction with water. ($Ca_3P_2 + 6H_2O \rightarrow 3Ca(OH)_2 + 2PH_3$)
Amides	$NaNH_2$	Violent reaction with water to yield ammonia. ($NaNH_2 + H_2O \rightarrow NH_3 + NaOH$)

Table 27-1. Summary of properties of reactive materials which can create hazardous conditions under special conditions.

melting points and low chemical reactivity. They are used as abrasive materials and for cutting edges of machine tools. Among those which are especially hard and inert are

- SiC – silicon carbide.

- TiC – titanium carbide.

- TaC – tantalum carbide.
- W_2C – tungsten carbide.

OTHER REACTIVE MATERIALS

Other reactive materials which may create hazardous conditions under certain conditions are listed in table 27-1.

REVIEW

1. Metal hydrides react with water to form

   a. Oxygen.
   b. Hydrogen.

   c. Acids.
   d. Chlorides.

2. Metal hydrides react with acids to form

   a. Hydrogen.
   b. Sodium chloride.

   c. Oxygen.
   d. Acids.

3. An example of a nonmetal hydride is

   a. Methane.
   b. Sodium hydride.

   c. Carbon tetrachloride.
   d. None of these.

4. All metal hydrides react with

   a. Oxidizing agents.
   b. Water.

   c. Both a and b.
   d. Neither a nor b.

5. Metal hydrides are

   a. Weaker reducing agents than nonmetal hydrides.
   b. More reactive with water than nonmetal hydrides.
   c. Lower boiling point materials than nonmetal hydrides.
   d. None of the above.

6. Nonmetal hydrides are usually

    a. Nontoxic.   c. Oxidizing agents.
    b. Toxic.   d. Solids.

7. Carbides are hazardous reactive materials which react with water to form

    a. Hydrogen.   c. Methane.
    b. Acetylene.   d. Benzene.

8. Which of the following is an example of a nonhazardous carbide?

    a. Be_2C.   c. CaC_2.
    b. Al_4C.   d. TiC.

9. Sodium sulfide is hazardous because it

    a. Forms sulfur dioxide when burned in air.
    b. Forms phosphine by reaction with water.
    c. Forms ammonia by reaction with bases.
    d. Forms hydrogen by reaction with water.

SECTION 7 CORROSIVE MATERIALS

Unit 28 Acids

Many acids such as hydrochloric, sulfuric, and nitric acids have the corrosive properties generally associated with the term "acid" in the minds of most people. Others like citric acid are common constituents of food and are weak, relatively noncorrosive acids.

The term acid signifies a certain type of chemical behavior. An acid is a compound which donates a proton to water. This is a fundamental definition of an acid. The hydronium ion, H_3O^+, is actually a hydrated proton (H^+).

$$HCl + H_2O \rightleftharpoons H_3O^+ + Cl^-$$

$$HNO_3 + H_2O \rightleftharpoons H_3O^+ + NO_3^-$$

The term hydrogen ion (the usual symbol is H^+) used by the chemist is identical in every respect to the term proton (the usual symbol is p) used by the physicist.

The characteristics of an acid are exhibited more strongly if the concentration of hydronium ion is greater. Hence, an acid is characterized as being stronger or weaker than another, according to whether a higher or lower concentration of hydronium ions is present in a solution containing a given total concentration of acid. Thus hydrochloric acid is called a strong acid because it is hydrolyzed almost completely to H_3O^+ by water. In contrast, acetic acid (vinegar) is a weak acid because it is only slightly hydrolyzed to H_3O^+ This definition provides a fundamental criterion for rating the hazardous properties of various acids as listed in table 28-1.

Water solutions of the acids at the top of table 28-1 have the properties of strong acids and the corrosive hazards are the greatest. By corrosive hazard is meant the hazard associated with the hydronium ion, H_3O^+. The corrosive hazards associated with H_3O^+ concentrations are

- Reaction with metals to liberate hydrogen and a base.

$$H_3O^+ + Na \rightarrow H_2 + NaOH$$

- Neutralization reactions which can be dangerously rapid.

$$H_3O^+ + OH^- \rightarrow 2H_2O$$

- Reaction with body constituents such as proteins.

In table 28-1 acids are rated as strong or weak acids. Some of the weak acids are not hazardous by virtue of acid strength but are highly hazardous for other reasons. Hydrogen cyanide (HCN) and hydrogen sulfide (H_2S) are weak acids but are highly toxic. Acid strength rates ability to generate H_3O^+ only; the other components of the acid molecule

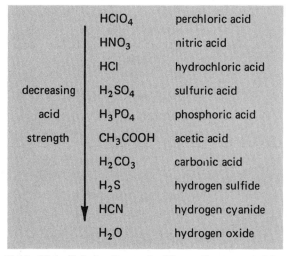

	$HClO_4$	perchloric acid
	HNO_3	nitric acid
	HCl	hydrochloric acid
decreasing	H_2SO_4	sulfuric acid
acid	H_3PO_4	phosphoric acid
strength	CH_3COOH	acetic acid
	H_2CO_3	carbonic acid
	H_2S	hydrogen sulfide
	HCN	hydrogen cyanide
	H_2O	hydrogen oxide

Table 28-1. Relative Strength of Some Common Acids.

Name	Formula	Acid Strength	Oxidizing Agent	Reducing Agent	Toxic	Water Reactive
hydrochloric acid	HCl	strong	–	–	+	–
hydrofluoric acid	HF	weak	–	–	+	–
hydrobromic acid	HBr	weak	–	–	+	–
hydriodic acid	HI	weak	–	–	+	–
sulfuric acid	H_2SO_4	strong	+	–	+	+
fuming sulfuric acid (oleum)	H_2SO_5	strong	+	–	+	+
nitric acid	HNO_3	strong	+	–	+	–
nitrous acid	HNO_2	strong	–	+	+	–
perchloric acid	$HClO_4$	strong	+	–	+	–
chloric acid	$HClO_3$	strong	+	–	+	–
chlorous acid	$HClO_2$	strong	–	+	+	–
hypochlorous acid	$HClO$	strong	–	+	+	–
phosphoric acid	H_3PO_4	strong	–	–	+	–
formic acid	HCOOH	weak	–	–	+	–
acetic acid	CH_3COOH	weak	–	–	+	–

Table 28-2. Hazardous Properties of Acids.

must be considered also with regard to hazardous properties as summarized in table 28-2.

- Acids can also be oxidizing agents. Thus, perchloric acid ($HClO_4$) is hazardous because it is both a strong acid and a strong oxidizing agent. Nitric acid (HNO_3) is another familiar example of a strong oxidizing acid.

- Acids can also be reducing agents. Nitrous acid (HNO_2) is an example of an acid which is strong and is also a reducing agent.

- Acids can also be toxic.

- Acids can be violently reactive with water. The exothermic reaction of sulfuric acid with water is a familiar example.

In addition to the primary acids listed in table 28-2, many materials which are not themselves acids can be converted to acids.

- The halogens (F_2, Cl_2, Br_2, I_2) react with water to form the corresponding acids.

- All oxides of nonmetals react with water to form acids.

$$SO_3 + H_2O \rightarrow H_2SO_4$$

$$NO_2 + H_2O \rightarrow HNO_3$$

$$CO_2 + H_2O \rightarrow H_2CO_3$$

- Organic compounds such as acid halides react with water to form acids.

COCl benzoyl chloride $+ H_2O$ COOH benzoic acid $+ HCl$

CH_2Cl benzyl chloride $+ H_2O$ CH_2OH benzyl alcohol $+ HCl$

The acid strengths of most of the organic acids can be deduced by analogy to the common inorganic acids listed in table 28-1.

ORGANIC ACID	INORGANIC ACID	ORGANIC ACID	INORGANIC ACID
Phenyl perchlorate ClO_4 (benzene ring)	Perchloric acid $HClO_4$	Benzoic acid $COOH$ (benzene ring)	Acetic acid CH_3COOH
Benzene sulfonic acid SO_3H (benzene ring)	Sulfuric acid H_2SO_4	Phenylmercaptan SH (benzene ring)	Hydrogen sulfide HSH
Phenylphosphinic acid PO_3H_2 (benzene ring)	Phosphoric acid H_3PO_4	Phenylnitrile CN (benzene ring)	Hydrogen cyanide HCN

HAZARDS OF THE ACIDS THE FIRE FIGHTER MAY ENCOUNTER

Some of the more common acids that fire fighters are apt to encounter are discussed here.

Sulfuric Acid (H_2SO_4)

Sulfuric acid, commonly called battery acid, dipping acid, or oil of vitriol, is not flammable, but it does present certain health hazards to the fire fighter as most of the inorganic acids do. Since it is corrosive, it causes deep and severe burns to the skin. Sulfuric acid differs from most of the other inorganic acids in that it has a great affinity for water. This presents an added problem in fire control since the addition of small amounts of water to an acid spill causes boiling of the water with steam explosions, followed by splattering. This type of reaction has prompted chemists to follow the basic rule: "Never add water to acid; add acid to water." This, however, is an impossible rule for the fire fighter to follow. Therefore, when an acid spill (such as sulfuric acid) must be diluted, large amounts of water must be added *initially,* with the fire fighters fully protected and at a safe distance away. Failure to follow these rules will result in severe chemical burns.

Oleum ($H_2S_2O_7$), or fuming sulfuric acid, is another form of sulfuric acid that is highly reactive with water. It is sometimes referred to as "super concentrated" sulfuric acid. It derives its name from the white fumes emitted on contact with moisture in the air. Both forms are strong oxidizers.

Oleum is far more hazardous than sulfuric acid. Its reaction with water is violent with explosions resulting. It is such a strong dehydrating and oxidizing agent that it has been known to ignite combustibles. The viscous liquid has a choking odor.

Both oleum and sulfuric acid are shipped in containers that vary in size from one gallon to tank cars. To ship diluted forms of sulfuric acid requires special steel containers because of the increased corrosiveness with water. It may also be shipped in polyethylene, polypropylene, or rubber lined containers.

The Halogen Acids

The halogen acids — hydrochloric (HCl), hydrobromic (HBr), hydrofluoric (HF), and hydroiodic (HI) — are nonflammable acids that are all corrosive, toxic, and extremely irritating when present as a gas in air. They are highly soluble in water and do not present as great a problem as sulfuric acid does when diluted with water. Hydrofluoric acid is unique and separate from the others because it will dissolve (etch) glass.

The halogen acids are given off as by-products of combustion in most of the flame retardant plastics. This type of hazard increases each year as more and more bromine-, chlorine-, and fluorine-containing plastics are used in making flame retardant materials.

Leaks of hydrogen fluoride gas should be handled with *extreme* caution and adjacent areas should be evacuated. Fire fighters should not approach the hazard areas unless a life is at stake. Protective clothing used in the clean-up of spills must be known to be impervious to these acids. The corrosive nature of these acids makes cleanup especially hazardous.

Hydrobromic, hydrochloric and hydro-iodic acids are shipped in glass containers or polyethylene or lined drums. Hydrofluoric acid is shipped in special steel or polyethylene containers.

Nitric Acid (HNO$_3$)

Commonly called aqua fortis, nitric acid is a colorless to brown liquid that is non-flammable. It is a strong oxidizer and has the ability to dissolve copper, which is untouched by other acids. Reddish fumes are given off by the acid. These fumes of nitrogen dioxide are toxic. Since nitric acid is a strong oxidizer it has been known to react explosively with metallic powders and natural organic compounds such as turpentine and carbides. Nitric acid is used in the manufacture of nitro compounds such as nitrobenzene and TNT.

Skin contact with nitric acid causes severe burns. The skin rapidly turns yellow within minutes after contact.

During fires or accidents involving nitric acid, a number of oxides of nitrogen may be present. These include nitrous oxide (N_2O), nitric oxide (NO), nitrogen trioxide (N_2O_3), and the highly toxic nitrogen dioxide (NO_2). Pulmonary damage caused from the inhalation of nitrogen dioxide, which converts to nitric acid in the lungs, is generally not apparent until four or more hours after exposure to the nitric acid fumes.

Nitrating Acid (Mixed Acid)

This mixture of nitric acid and sulfuric acid is extremely corrosive. It is a colorless to light yellow liquid of varying proportions. Like nitric acid, the mixture is not flammable but may cause spontaneous fires on contact with combustibles, such as wood, cotton or chemicals, and certain solvents and organic fuels.

Since the mixture is so corrosive, spills should be flushed with water, but water must not be allowed to enter any tank that holds the liquid.

Perchloric Acid (HClO$_4$)

Perchloric acid is so reactive that the pure form of it is highly unstable. The pure liquid decomposes and explodes spontaneously. Water slows down the decomposition of this oily, colorless liquid. Solutions of 60- to 72-percent perchloric acid are commonly found in laboratories in one pound reagent bottles. Storage of larger amounts of this liquid (carboys) must be kept away from all organic combustibles and dehydrating agents. Strong dehydrating agents can convert the water solution to the unstable and highly reactive anhydrous form.

As perchloric acid is warmed, it becomes an extremely strong oxidizing agent. In this warm stage, explosive conditions occur — either spontaneously or by continued heating.

Fires around perchloric acid should be fought with great caution. The combination of heat and dehydration make explosions common. Large amounts of water for cooling and dilution are needed. Since perchloric acid is similar to sulfuric acid in evolution of heat when water is applied, large initial amounts of water must be used when handling perchloric spills.

Perchloric acid is an acute irritant to eyes, skin, and mucous membrane. Remember — clothing that comes in contact with perchloric becomes extremely sensitive to shock.

Organic Acids

The organic acids present an additional problem to fire fighters — that of flammability. Like most carbon-containing compounds, degree of flammability varies with the type of acid in question.

Acetic Acid (CH_3COOH)

Acetic acid is found in two forms, glacial (concentrated) and diluted. Solutions diluted to 4-6-percent acetic acid are known as vinegar.

Glacial acetic acid is a colorless combustible liquid with a flash point of 109°F. It has a flammable range between 4 and 16 percent. It causes severe burns to the skin and the eyes. The addition of water to spills greatly reduces the dangers of this acid.

The anhydrous form, acetic anhydride, differs from the glacial form in that it is highly reactive with water. It reacts vigorously with water to form acetic acid. Acetic anhydride not only affects the eyes and skin but also is a respiratory irritant. In fighting fires, if water is to be used, large amounts from master streams are suggested.

Formic Acid ($HCOOH$)

Formic acid is a combustible liquid with a flash point of 156°F and flammable limits of 18 to 57 percent. The commercial, technical form (90%) in water has a flash point of 122°F.

Those who have been bitten by ants know the irritating sting that lingers. This is caused by formic acid. It is an acute irritant that is highly toxic even if inhaled on very short exposures. The liquid burns the skin and eyes. Most fire fighters are familiar with formic acid from entering fire areas and feeling their eyes burn. Formic acid is a by-product of burning wood.

Water is used to fight fires or flush spills. The acid is shipped as a corrosive liquid in stainless steel containers.

Phenol

Phenol is an extremely dangerous Class B poison with a flash point of 175°F. Its odor is typical of a disinfectant. Phenol has a low toxicity limit (5 ppm). The greatest hazard of phenol is that it causes severe burns to the skin. A small amount on the skin is enough to cause serious illness or even death. Contact with it causes the skin to become reddened. Prolonged contact makes the skin white and wrinkled. There is no initial sensation of pain for phenol is an anesthetic. Pain becomes evident as the chemical works its way into the skin.

Extreme care should be taken when near any fires or spills involving phenol. Any fire fighter who comes in skin contact with phenol should flush the contaminated area with water and then have immediate medical assistance. He must not neglect to report and treat any phenol burn no matter how small it may be.

Phenol spills (it melts at 105°F) or solid phenol can be flushed away. However, extreme care should be taken with approved resistant protective gear.

Phenol is shipped in the pure form or it may be shipped in the crude form. It comes in small reagent bottles, or it may be shipped in tank wagon, tank cars, or barges.

REVIEW

1. An acid is a compound which
   a. Dissolves metals.
   b. Forms a hydronium ion in water.
   c. Forms a hydroxyl ion in water.
   d. Forms hydrogen in water.

2. A hydronium ion is

   a. H_2.   c. OH^-.

   b. H.   d. H_3O^+.

3. Which of the following is a weak acid?

   a. Sulfuric acid.   c. Sodium hypochlorite.

   b. Benzoic acid.   d. Nitrous acid.

4. All strong acids are

   a. Corrosive hazardous materials.   c. Explosive hazardous materials.

   b. Toxic hazardous materials.   d. All of these.

5. All weak acids are

   a. Corrosive hazardous materials.   c. Explosive hazardous materials.

   b. Toxic hazardous materials.   d. None of these.

6. Which of the following acids reacts exothermically with water?

   a. HCl.   c. HNO_3.

   b. H_2SO_4.   d. None of these.

7. Which of the following acids is also an oxidizing hazardous material?

   a. Sulfuric acid.   c. Formic acid.

   b. Phosphoric acid.   d. Hydrobromic acid.

8. Which of the following acids is also a hazardous reducing agent?

   a. Nitrous acid.   c. Hydrochloric acid.

   b. Phosphoric acid.   d. Hydrobromic acid.

9. Which of the following acids is also a toxic hazardous material?

   a. Hydrochloric (HCl).   c. Hydrofluoric (HF).

   b. Hydrogen sulfide (H_2S).   d. All of these.

10. Which of the following is the strongest acid?

   a. Phosphoric (H_3PO_4).   c. Sulfuric (H_2SO_4).

   b. Nitric (HNO_3).   d. Perchloric ($HClO_4$).

Unit 29 Bases

Bases such as sodium hydroxide and ammonia have the corrosive properties generally associated with the term *base* in the minds of most people. A base is a compound which forms a hydroxide ion (OH⁻) in water solution.

$$NH_3 + H_2O \rightleftharpoons NH_4^+ + OH^-$$
$$Na_2O + H_2O \rightleftharpoons 2Na^+ + 2OH^-$$

A strong base is a compound which yields a high concentration of hydroxide ion; a weak base yields a low concentration of hydroxide ion. The relative strengths of the bases are shown in table 29-1.

The corrosive hazards associated with OH⁻ concentration are

- Neutralization reactions which can be dangerously rapid.

- Reaction with body constituents such as proteins.

The strongest bases are amines such as methylamine and ethylamine. Bases such as sodium hydroxide, although extremely basic, are not as strong as the amines. The carbonates are weak bases.

As with the acids, the other components of the base molecule must be considered also.

- All amines are reducing agents.

- Amines and solid metal hydroxides can be violently reactive with water. (The exothermic reaction of sodium hydroxide (lye) with water is a familiar example.)

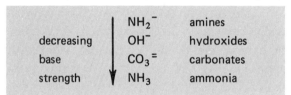

Table 29-1. Relative Strengths
of Some Common Bases.

- All amines liberate the toxic gas, ammonia, in the presence of bases. For example, dry foam extinguishers liberate ammonia (NH_3) from an amine.

$$CH_3NH_2 \quad \rightarrow \quad NH_3$$
any base

- All amines are toxic. The toxicity of metal hydroxides is determined by the metal rather than the hydroxide group (OH).

As was the case with acids, the hazardous properties of bases are determined by factors other than the base strength. The hazardous properties of several typical bases are given in table 29-2. All of the bases such as the amines, hydroxides, and ammonia tend to be severe corrosive and toxic hazards. In contrast, the carbonates which are weak bases are essentially nonhazardous except for the possible indirect hazard associated with the metal cation. For example, calcium carbonate is nonhazardous but beryllium carbonate is a highly toxic material because of the beryllium cation.

HAZARDS OF INORGANIC BASES APT TO BE ENCOUNTERED BY FIRE FIGHTERS

Like acids, bases are divided into two groups; inorganic and organic. The inorganic bases are nonflammable while the organic bases, containing carbon and hydrogen, are flammable. The health hazards of both are essentially moderate, with the organic amines carrying a greater degree of toxicity than the inorganic.

Sodium Hydroxide

Sodium hydroxide, commonly called caustic soda or lye, is the most common com-

Name	Formula	Base Strength	Reducing Agent	Toxic	Water Reactive
methyl amine (all organic amines)	CH_3NH_2	high	+	+	+
aniline	NH_2 (benzene ring)	high	+	+	+
all metal hydroxides	OH^-	high	-	+	+
all metal carbonates	$CO_3^=$	low	-	-	-
ammonia	NH_3	low	+	+	+

Table 29-2. Hazardous Properties of Bases.

mercial caustic. It is found as white deliquescent flakes, sticks, beads, lumps, or as a solid mass. In the liquid form, it is found usually dissolved in either water or alcohol at concentrations of 50 to 75 percent sodium hydroxide. The solution form is viscous at this high concentration.

The solid form of sodium hydroxide absorbs water at such a great degree that heat is given off in the same manner as with sulfuric acid. In some cases, there is enough heat to char and even ignite combustible materials. Contact with some metals, aluminum for example, will generate hydrogen gas, thus increasing potential for a fire or explosion.

Sodium hydroxide is toxic and presents a severe hazard to skin tissue. Contact with it causes severe burns if immediate medical attention is not given. It is highly toxic by ingestion for the same reason.

Sodium hydroxide is used widely throughout the chemical industry. It can be found in such areas as pulp and paper mills, textile processing, petroleum refining, chemical and soap manufacturing, medicine, etching and electroplating areas, and in chemical laboratories.

Caustic soda is shipped in one-pound bottles and in drums and barrels; in solution, it is shipped as a white label, corrosive liquid in quantities from one-pound bottles to barge and tank car loads.

Liquid spills should be flooded with copious amounts of water and diked if possible. No fire fighter should enter the liquid area. Full protective clothing (rubberized if possible) should be worn.

Caustic Potash

Caustic potash is very similar in all aspects to caustic soda except for the formula. It is used widely in chemical labs and plants for the manufacture of soap, bleaches, medicine, and other potassium salt substances. Any accidents involving caustic potash should be treated in the same manner as incidents involving caustic soda.

Ammonium Hydroxide

This familiar household ammonia is made by bubbling ammonia gas through water.

$$NH_3 + H_2O \rightarrow NH_4OH$$

Synonyms for this solution are aqua ammonia, ammonium hydrate, and ammonia solution. Concentrations of ammonia gas in water generally go up to 30 percent.

Ammonium hydroxide solutions are generally shipped in steel tank trucks, tank cars, or barges. Large industrial users begin with ammonia gas and convert it to ammonium hydroxide.

The pungent odor of ammonia is very irritating. It is highly toxic by ingestion and extremely irritating to the eyes. Fire fighters should be aware of the pungent odor and toxicity of this solution. It is only slightly basic as compared to the strong bases, such as caustic potash or soda, and does not cause the burns that the caustic potash and caustic soda do.

HAZARDS OF COMMONLY ENCOUNTERED ORGANIC BASES

Organic bases are not only highly toxic when compared to the inorganic bases, but are also flammable. The degree of flammability varies, but this combustible nature produces even more toxic fumes (HCN) from the decomposition of the amine bases.

The organic bases are derived from organic functional groups called amines. Amines have the nitrogen groups which identify them.

The fire fighter should be aware of any label that carries the term *amine*. Most of the amines are highly toxic, are either flammable or combustible, and require special attention. They are all classified as either flammable or combustible liquids or solids or gases. All

organic amines are highly toxic by absorption through the skin.

Aniline

Aniline is perhaps the most important of the organic bases. It is a liquid, colorless to dark brown in color. It has a flash point of 158°F and a boiling point of 364°F. It has been used for years in the manufacture of dyes and drugs.

Like all organic amines, aniline is highly toxic. Aniline, or aniline oil, aminobenzene or phenyl amine as it is sometimes called, is classified as a Class B poison. It is highly toxic by absorption through the skin. If the skin or eyes are affected, they must be flushed immediately with plenty of soap and water for at least fifteen minutes. Speed is of utmost importance in removing this poison because aniline is readily absorbed through the skin.

Fires involving aniline should be handled using full protective clothing and self-contained breathing apparatus. Fires should be extinguished with water spray or alcohol foam. Vapor can be dispersed using fog.

	DIETHYLENETRIAMINE	METHYLAMINES
Synonyms	Diaminodiethylamine	Mono, di, trimethylamine
Solid, Liq. - Gas	Corrosive liquid	Liquid
Flash point of (°F)	215	Gas
Flam. range	NA	5-21
Soluble in water	Yes	Yes
Vapor density	3.5	1 to 2
Health hazard	Toxic — contact with tissue or eyes will cause burns.	Liquid causes burns. Vapor irritating. Toxic liquefied gas.
Spill or leak	Evacuate area. Avoid contact with liquid or vapor. Vapor density 3.5. Wear protective clothing and self-contained breathing apparatus.	Evacuate area. Avoid contact with liquid or vapor — toxic liquefied gas. Wear protective clothing and self-contained breathing apparatus.
Fire	Water, dry chemical, CO_2	Water, dry chemical, CO_2

Table 29-3. Other Organic Bases.

Diethylamine

Diethylamine is a colorless flammable liquid. It has a flash point below 0°F. The odor of diethylamine is that of ammonia but also — characteristic of most aliphatic amines — fishy.

The liquid causes tissue burn and is a strong irritant. The vapor is toxic and causes lung injury. It is moderately toxic by ingestion.

Diethylamine has a vapor density of 2.5, which is extremely heavy. This, along with a flash point below 0°F, makes it a high fire hazard if a spill occurs. The flammable limits for this liquid are between 1.8 to 10.1.

Large fires can be extinguished using water. Full protective clothing and breathing apparatus should be worn at all times since the vapors are toxic. Carbon dioxide and dry chemical can also be used.

Spills can be flushed with water or vapors can be dispersed in air with fog. An important point to keep in mind is the vapor density of the vapors. They can travel a long distance — then flashing back, they can cause either a fire or an explosion. Flammable regions, therefore, should only be entered for the purpose of rescue.

Other

Properties and hazards of other bases are shown in table 29-3.

REVIEW

1. Sodium hydroxide is a base because it reacts with water to form a

   a. Hydronium ion.   c. Proton.
   b. Hydroxide ion.   d. Neutron.

2. Ammonia is a base because

   a. It reacts with water to form a hydroxide ion.
   b. It reacts with water to form an ammonium ion.
   c. It reacts with water to form ammonium chloride.
   d. All of the above.

3. All amines are

   a. Bases.   c. Reactive with acids.
   b. Reducing agents.   d. All of these.

4. A base is

   a. An alkaline material.   c. An acidic material.
   b. A neutral material.   d. An explosive material.

5. Sodium carbonate is

   a. A neutral material.   c. A weak base.
   b. A strong base.   d. A weak acid.

6. All amines react with sodium hydroxide to form

   a. Ammonia.   c. Amine hydrates.
   b. Water.   d. Neutral products.

7. The toxicity of metal hydroxides is determined by
    a. The hydroxide content.   c. The amine.
    b. The metal.   d. The base strength.

8. Which of the following is the strongest base?
    a. Sodium hydroxide.   c. Ammonium hydroxide.
    b. Calcium carbonate.   d. Ethanolamine.

9. Which of the following bases is a reducing agent?
    a. Aniline.   c. Ethyl amine.
    b. Ammonia.   d. All are reducing agents.

10. Which of the following compounds is a hazardous material?
    a. Beryllium carbonate.   c. Potassium carbonate.
    b. Calcium carbonate.   d. All of these.

TOXIC MATERIALS

Unit 30 Principles of Toxicology

In order to avoid or minimize injury from toxic materials it is necessary to understand how the materials enter the body and how they function as poisons within the body. Although no single method of classifying toxic materials is entirely satisfactory, considerable insight is obtained by classifying them in terms of their physiological action.

IRRITANTS

Irritants are corrosive materials which attack the mucous membrane surfaces of the body. All are highly reactive chemicals which react quickly with body tissues.

• Materials which are very water soluble tend to be removed in the moist upper respiratory tract and primarily attack the upper respiratory tract. Some examples are listed:

Hydrochloric acid	HCl
Hydrofluoric acid	HF
Sulfur dioxide	SO_2

• Materials which are only moderately soluble in water tend to be partially removed in the upper respiratory tract, but some of the material penetrates into the lungs. These materials attack both the upper respiratory tract and the lungs. Some examples are listed:

Halogens	F_2, Cl_2, Br_2
Ozone	O_3
Phosphorus trichloride	PCl_3
Phosphorus pentachloride	PCl_5

• Materials which are only slightly soluble in water penetrate into the far reaches of the lungs and primarily attack the terminal respiratory passages called the alveoli. Some examples are listed:

Nitric oxide	NO
Nitrogen tetroxide	N_2O_4
Phosgene	$COCl_2$

ASPHYXIANTS

Asphyxiants interfere with the oxidation processes in the body.

• The **simple asphyxiants** are physiologically inert gases which dilute or replace the oxygen required for breathing. The simple asphyxiants are often called **suffocants**. Some examples are listed:

Carbon dioxide	CO_2
Helium	He
Hydrogen	H_2
Nitrogen	N_2
Saturated hydrocarbons like methane and ethane	

• The **chemical asphyxiants** react with an essential body function involved with transportation of oxygen from the lungs via the red blood cells to the body tissues. Asphyxiation results even though adequate oxygen is present. The chemical asphyxiants are often called **respiratory poisons**. Some examples are listed:

Carbon monoxide	CO
Hydrogen cyanide	HCN
Aniline	NH_2–⬡
Nitrobenzene	NO_2–⬡
Sodium nitrite	$NaNO_2$
Hydrogen sulfide	H_2S

ANESTHETICS AND NARCOTICS

The anesthetics and narcotics are hazardous materials which depress the central nervous system and lead to unconsciousness. Some examples are listed:

Acetylene	$CH \equiv CH$
Ethylene	$CH_2 = CH_2$
Diethyl ether	$CH_3 CH_2 OCH_2 CH_3$
Acetone	$CH_3 \overset{\overset{\displaystyle O}{\|\|}}{C} CH_3$
Ethyl alcohol	$CH_3 CH_2 OH$

SYSTEMIC POISONS

Systemic poisons are hazardous materials which injure or destroy internal organs of the body.

• Liver poisons may be fat soluble materials and tend to be concentrated in the liver. Some examples are listed:

All halogenated hydrocarbons

Benzene

Toluene $- CH_3$

Xylene $H_3C - \bigcirc - CH_3$

Naphthalene

• Liver poisons may be toxic metals. This is usually referred to as heavy metal poisoning. Some examples are listed:

Lead	Pb
Mercury	Hg
Beryllium	Be

• Liver poisons may be toxic nonmetals. Although not chemically correct, this is also commonly referred to as heavy metal poisoning. Some examples are listed:

Arsenic	As
Sodium fluoride	NaF

• The nerve poisons function by blocking essential nerve impulses throughout the body. Some examples are listed:

Organic phosphates	
Carbon disulfide	CS_2
Methyl alcohol	$CH_3 OH$

ABSORPTION OF TOXIC MATERIALS

The three primary routes of absorption of hazardous toxic materials into the body are shown:

• Absorption through the skin which is the most important route for the hydrocarbons, halogenated hydrocarbons and organic amines.

• Gastrointestinal absorption which most often involves the accidental swallowing of harmful amounts of toxic materials.

• Absorption through the lungs which accounts for the vast majority of hazardous toxic materials.

Although the physiological response of the human body to many types of hazardous toxic materials is fairly well understood, the so-called safe limit for various exposure times and conditions is at best an "educated guess." The basic reason for uncertainty in much of the toxicity data available relates to variability among test subjects.

• Susceptibility to allergies.

• Previous history of illness.

• Body weight.

• Inherent difference in the body chemistry of each individual.

All toxicity data must be regarded as an estimate. Tables of numerical data such as are shown in following units are much more helpful than the broad classifications A, B, and C poisons such as are used by many agencies. Both types of data are used in the discussions that follow. The essential point to remember in dealing with toxicity data is that all such data is approximate and should be used as a guide only.

TOXICITY LIMITS

In the United States, the threshold toxicity limits set by the American Conference of Governmental Industrial Hygienists are the most universally accepted quantitative toxicity criteria. These toxicity values are usually referred to as *ACGIH accepted values.* All threshold limits (TL) listed in this textbook are ACGIH accepted TL values and as such are typical of the best values available at this time. These data are revised annually, so for critical applications the reader is strongly advised to refer to the most recent official publication of the American Conference of Governmental Industrial Hygienists.

For toxic gases, the TL values are given in parts per million (ppm) of the toxic material per million parts of air. Thus 1 ppm is 1 part in 1,000,000 parts of air.

For finely divided dusts, particulate matter, and aerosols the TL value is expressed in milligrams of solid per cubic meter of air. This unit of measurement is used because it is more feasible to weigh solids suspended in air than to try to measure the volume.

For some dusts such as silicon dioxide and asbestos, the TL is expressed in millions of particles per cubic foot of air. This convention is based on common usage rather than scientific necessity or logic.

Another widely used criteria for toxicity rating is that of the United States Department of Transportation (DOT). (The official DOT and other related safety criteria are discussed in detail in unit 37.) This is a qualitative rating system which clearly identifies the relative hazard and type of hazard associated with various types of materials. Hazards are identified by a simplified label system.

- Class A — extremely dangerous poisons.
- Class B — less dangerous poisons.
- Class C — irritants and tear gases.
- Class D — radioactive materials.
- Corrosive liquids — no class identification because in the sense of strict definition they are not poisons.

REVIEW

1. An irritant is a corrosive material which attacks the
   - a. Skin.
   - b. Upper respiratory tract.
   - c. Lower respiratory tract.
   - d. Any of these.

2. An example of an irritant is
   - a. Sulfur dioxide.
   - b. An acid.
   - c. A base.
   - d. Any of these.

3. Which of the following is an irritant?
   - a. An acid chloride.
   - b. Hexane.
   - c. Carbon monoxide.
   - d. None of these.

4. Irritants which are only slightly water soluble are
   - a. Lower respiratory poisons.
   - b. Not hazardous.
   - c. Upper respiratory poisons.
   - d. Acids.

5. Asphyxiants are hazardous because they

   a. Are water soluble.
   b. Are gases.
   c. Interfere with oxidation processes in the body.
   d. Have vapor densities greater than air.

6. Which of the following is a simple asphyxiant?

   a. Air.   c. Oxygen.
   b. Hydrogen.   d. None of these.

7. A chemical asphyxiant differs from a simple asphyxiant because

   a. It has a complex chemical formula like $NaNO_2$.
   b. It reacts chemically with the human body.
   c. It reacts with a portion of the body system involved with the trans-portation of oxygen by the blood.
   d. It is an irritant as well as an asphyxiant.

8. Asphyxiation results when

   a. Insufficient oxygen reaches the body cells.
   b. The upper flammable limit is exceeded.
   c. The 80 percent nitrogen component of the air is replaced by an equivalent amount of helium.
   d. Any of the above.

9. Which of the following is a simple asphyxiant?

   a. CO.   c. HCN.
   b. $H_2 S$.   d. N_2.

10. Acetone is a toxic hazardous material because it is

    a. An irritant.   c. A respiratory poison.
    b. A narcotic.   d. A liver poison.

11. Which of the following types of toxic hazardous materials are most alike?

    a. Nerve and liver poisons.
    b. Systemic and respiratory poisons.
    c. Anesthetic and narcotic poisons.
    d. Irritant and heavy metal poisons.

12. Which of the following are systemic poisons?

    a. Heavy metals.   c. Liver poisons.
    b. Nerve poisons.   d. All of these.

13. Toxic hazardous materials can be absorbed through the

    a. Lungs.   c. Gallbladder.
    b. Nerves.   d. Bones.

14. A toxic hazardous material which is ingested is most likely to be absorbed through the

    a. Lungs.   c. Gastrointestinal system.
    b. Skin.   d. None of these.

15. Toxicity data must be regarded as approximate because

    a. Some humans have allergies.
    b. All humans do not have the same weight.
    c. There is a significant difference in the body chemistry of each individual.
    d. All of the above.

Unit 31 Respiratory Poisons

Everyone is familiar with the respiratory process to a certain extent. Oxygen which is present in the atmosphere at a concentration level of 21 percent must be transported to the lungs via the bloodstream to the cells so that they can continue to live. Without oxygen, the brain cells perish in three to five minutes. The cells of the heart and central nervous system are particularly sensitive to low levels of oxygen in the blood because of their essential functions.

- Anoxia means complete absence of oxygen in the blood.

- Hypoxia means low levels of oxygen in the blood.

- Hyperoxia means excessive levels of oxygen in the blood.

Respiration is usually subdivided further to take into account the dual process in which oxygen is inhaled and carbon dioxide is exhaled.

- External respiration in which oxygen enters the blood and carbon dioxide leaves the lungs at the lung interfaces.

- Internal respiration in which oxygen leaves the blood to be taken up by the individual cells and carbon dioxide waste products from the cells are picked up by the blood.

From the point of view of toxicology the respiratory system is most conveniently subdivided into the upper and lower respiratory system. Consideration of the upper and lower respiratory tracts as well as the blood is an adequate physiological basis for understanding the effects of the common respiratory poisons.

- The upper respiratory tract consists of the nose, pharynx, larynx, and trachea.

- The lower respiratory tract consists of the lung structures including the bronchi, bronchioli, and alveoli.

- The oxygen-carrying component of the blood is the hemoglobin molecule.

The major components of the respiratory system are shown in figure 31-1. During inspiration, air enters the nose and then the nasal portion of the pharynx. Here the mucous secreted by the nasal mucous membranes traps the dust, pollen, and airborne bacteria to which the human body is constantly exposed. The magnitude and importance of this function is usually underestimated. Approximately a quart of mucous is produced daily by an adult to accomplish this defensive body function. This constant flow of mucous is swallowed with the saliva or in the case of nasal congestion is expelled through the nose. This

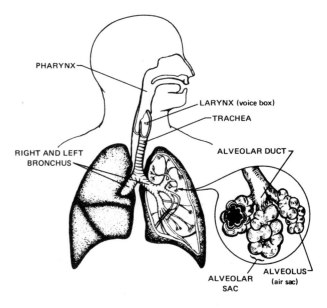

Fig. 31-1. The Human Respiratory Tract.

Name	Formula	Application	Threshold Toxicity Limit	
			ppm for gases	mgm per cubic meter in air for solids
ethyl silicate	$(C_2H_5)_4SiO_4$	masonry sealant	100	850
hydrogen bromide	HBr	chemical reagent	3	10
hydrogen chloride	HCl	concrete cleaner	5	7
hydrogen fluoride	HF	glass etchant	3	2
nitrogen dioxide	NO_2	- - -	5	9
sulfur dioxide	SO_2	toxic combustion product	5	13

Table 31-1. Toxicity of Upper Respiratory Poisons.

is also the point at which the upper respiratory poisons attack the human body. It also explains why upper respiratory poisons enter the stomach, proceed through the digestive tract, and are ultimately absorbed into the bloodstream to exert their effects in various secondary ways.

The inhaled air then proceeds through the remainder of the upper respiratory tract via the oral pharynx, larynx, and trachea to the bronchial tubes. Upper respiratory poisons usually show their effects in this area by hoarseness and sore throat.

UPPER RESPIRATORY POISONS

As a class, upper respiratory poisons are those which are soluble in water or are highly reactive with water. They are often referred to as irritants or respiratory irritants also. Threshold toxicity limits for some of the common upper respiratory poisons are given in table 31-1.

The Halogens

The three halogen acids HBr, HCl, and HF are gases which are highly soluble in water and are therefore extracted in the upper respiratory tract. The differences in the threshold toxicity limits result from the differing toxicity of the F, Cl, and Br atoms.

The Dioxides

Nitrogen dioxide and sulfur dioxide react rapidly with water with the formation of the corresponding acids, nitric acid and sulfurous acid. Sulfurous acid is a reducing acid and is rapidly oxidized by oxygen to sulfuric acid.

$$NO_2 + H_2O \rightarrow HNO_3$$
$$SO_2 + H_2O \rightarrow H_2SO_3$$
$$2H_2SO_3 + O_2 \rightarrow 2H_2SO_4$$

The acids such as nitric, sulfurous, sulfuric and most others are usually classified as corrosives because they are most likely to be encountered in accidental spills rather than inhalation of mists. On the other hand, whenever a situation is encountered in which the acid mist is breathed, the corrosive action is exerted in the respiratory tract and they function as respiratory poisons.

Ethyl silicate is a widely used masonry sealant which is often used without adequate respiratory protection. It is likely to be encountered in homes for that reason. It hydrolyzes to form silicic acid (H_4SiO_4). Ethyl silicate is less toxic than the other respiratory poisons discussed above because silicic acid is a relatively weak acid.

Name	Formula	Application	Threshold Toxicity Limit	
			ppm for gases	mgm per cubic meter of air for solids
acrolein	$CH_2{=}CHCHO$	toxic combustion product	0.1	0.25
acrylonitrile	$CH_2{=}CHCN$	polymerization monomer	20	45
bromine	Br_2	chemical reagent	0.1	0.7
chlorine	Cl_2	disinfectant	1	3
fluorine	F_2	chemical reagent	1	2
chloropicrin	CCl_3NO_2	explosive	0.1	0.7
epichlorohydrin	$CH_2{-}CH{-}CH_2Cl$ with O bridge	polymerization monomer	5	19
hydrogen cyanide	HCN	toxic combustion product	10	11
hydrogen sulfide	H_2S	sewer gas	10	15
methyl cyanide	CH_3CN	- - -	40	70
methyl mercaptan	CH_3SH	petroleum refineries	0.5	1
phosgene	$COCl_2$	toxic combustion product	0.05	0.2
phosphorus pentasulfide	PS_5	matches	- - -	1

Table 31-2. Toxicity of Lower Respiratory Poisons.

LOWER RESPIRATORY POISONS

The lower respiratory poisons listed in table 31-2 function in the same basic way as the upper respiratory poisons just discussed. They react more slowly with water and tend to be absorbed in the alveoli of the lungs. Many of the war gases were selected on the basis of this property. Phosgene, chlorine, and chloropicrin which were candidates as war gases hardly affect the upper respiratory tract but hydrolyze to yield HCl in the alveoli of the lungs. Phosgene is particularly toxic because it yields both HCl and CO. Chloropicrin and chlorine also form multiple toxic products. This behavior is reflected in their high threshold toxicity limits.

The Cyanides

The cyanides are another general class of lower respiratory poisons. Hydrogen cyanide is not corrosive. It does not react chemically with the water of the mucous membranes. It dissolves in the water and is transported via the bloodstream to the individual cells of the body where it blocks oxygen uptake at the cellular level by combining with the enzymes which control cellular oxidation. Oxygen uptake at the cellular level is blocked only as long as the cyanide is present. Normal cellular oxygen uptake resumes if death of the cells has not already occurred. The cyanides are particularly hazardous because of their low threshold toxicity level coupled with the fact that they are odorless.

Hydrogen cyanide is a toxic gas formed during combustion of nitrogen compounds. It is also formed by the hydrolysis of organic cyanides such as methyl cyanide and acrylonitrile. This hydrolysis occurs in the alveoli of the lungs when the organic cyanide compound is inhaled. It would be equally toxic if ingested and the HCN were formed in the stomach.

Inorganic cyanides such as sodium cyanide or calcium cyanide liberate gaseous HCN with acids but not with water. Thus, sodium cyanide is highly toxic but is not likely to be encountered as a respiratory poison because it is a nonvolatile solid. Contact with an acid, however, converts it to HCN which is a hazardous respiratory poison.

Inorganic cyanides are extensively used in electroplating baths. In this application they are usually used in alkaline solution (a base such as sodium or potassium hydroxide is added). This affords an added degree of protection. On the other hand, alkaline metal cyanide mists are highly toxic because HCN would form in the lungs.

The Sulfides

Hydrogen sulfide is also a powerful lower respiratory poison. Its mode of action is very similar to that of HCN. It is transported to the cells in solution but does not react with the hemoglobin of the blood as does carbon monoxide. It kills the cells by paralysis of the respiratory centers. The rate of death from H_2S is about the same as with HCN.

Organic sulfides such as methyl mercaptan yield H_2S by hydrolysis with water. Similarly, metal sulfides such as sodium sulfide or iron sulfide yield H_2S with acid but not with water.

Acrolein

Acrolein is more properly classified as a lower respiratory poison than an irritant. Although it is a lachrymator and is highly irritating to the respiratory tract, it is listed here rather than with other aldehyde irritants in table 31-1. It also causes asthmatic reactions presumably because of initiating bronchial spasms. Acrolein is important because of its occurrence as a toxic combustion product.

HEMOGLOBIN MOLECULE POISONS

Another large class of hazardous respiratory poisons react with the hemoglobin molecule in the blood in various ways. They are classified as lower respiratory poisons because they must be absorbed through the alveoli of the lungs before they act on the body. Carbon monoxide is the best known example of the lower respiratory poisons which act by affecting the oxygen-carrying capacity of the blood. Carbon monoxide is 200 times more strongly bonded to hemoglobin than oxygen. In this way the hemoglobin is prevented from carrying oxygen and cellular asphyxiation results. The toxic compounds listed in table 31-3 function in a similar way although the detailed mechanisms vary.

• Some toxic compounds irreversibly react with the hemoglobin molecule. This condition results in anoxia and ultimately asphyxiation. A well-known example is carbon monoxide.

• Some toxic compounds cause hemolysis of the red blood cells. This condition results in anoxia and ultimately asphyxiation. Examples are hydrazine, phenylhydrazine, and aniline.

• Some compounds cause complex and poorly defined malfunctions of the oxygen-carrying capacity of the blood. Examples are nitro compounds and aromatic hydrocarbons such as benzene and toluene.

Name	Formula	Application	Threshold Toxicity Limit	
			ppm for gases	mgm per cubic meter of air for solids
aniline	NH$_2$ (benzene ring)	chemical intermediate	5	19
benzene	(benzene ring)	solvent	10	30
carbon monoxide	CO	toxic combustion product	50	55
2, 4-dinitrotoluene	CH$_3$ (benzene ring) NO$_2$ NO$_2$	explosive	- - - -	1.5
hydrazine	NH$_2$NH$_2$	rocket propellant, chemical intermediate	1	1.3
phenylhydrazine	NHNH$_2$ (benzene ring)	reagent commonly used in chemical laboratories	5	22
nitrobenzene	NO$_2$ (benzene ring)	solvent	1	5
nitromethane	CH$_3$NO$_2$	fuel, solvent	100	250
nitroglycerin	CH$_2$ONO$_2$ CH$_2$ONO$_2$	explosive	0.2	2
tetranitromethane	C(NO$_2$)$_4$	explosive	1	8
toluene	CH$_3$ (benzene ring)	solvent	100	375
trinitrotoluene	CH$_3$ O$_2$N (benzene ring) NO$_2$ NO$_2$	explosive	0.2	1.5

Table 31-3. Toxicity of Respiratory Poisons Which Affect Oxygen-Carrying Capacity of Blood.

INHALATION PROBLEMS

Inhalation of dust and particulate matter is also a serious lower respiratory problem. Moderate amounts of particulate matter are removed from the lung by the normal process of mucous secretion and ordinarily do not cause respiratory difficulties. Above about 20 million particles per cubic foot of air most substances begin to present a serious problem because the lung cannot efficiently remove them. Typical examples are given in table 31-4. They are encapsulated by the lung tissue

Name	Formula	Occurrence	Threshold Toxicity Limit million particles per cubic foot of air
asbestos	- - -	industrial dust	5
alumina	Al_2O_3	industrial dust	50
beryllium oxide	BeO	industrial ceramics metal fires	0.002 mgm per cubic meter of air
beryllium	Be	metal machining	0.002 mgm per cubic meter of air
coal	- - -	mining	50
mica	- - -	industrial dust	20
silica	SiO_2	industrial dust	20
soot	C	combustion by-product	20

Table 31-4. Toxicity of Respiratory Poisons: Dust and Particulate Matter.

and extensive scar tissue can result. This condition is called fibrosis and impaired oxygen transfer across the lung membrane results from the buildup of scar tissue. Some of the many types of dust which fall into this category are household dust, asbestos, soot, mica, and silicon dioxide.

Inhalation of dusts containing metals or metal oxides can lead to particularly severe respiratory difficulties. Inhalation of beryllium oxide particles leads to a condition known as beryllosis. The beryllium oxide particles are encapsulated in fibrous tissue just as is the case with any other particle. The beryllium is very toxic and destroys adjacent tissue, thus providing a mechanism for continued scar growth. Many other metals and metal salts behave in a similar fashion.

RESPIRATORY POISON SYMPTOMS

Clinical and physiological symptoms of most of the respiratory poisons discussed here are given in table 31-5. In order to understand this table, it may be necessary to define many of the new words. The same vocabulary is used universally in all toxicity handbooks and compilations. Some physiology terms which may be new to the student are defined:

- Anoxemia — condition resulting from lack of oxygen in the blood.
- Bronchitis — inflammation of the bronchial tubes.
- Conjunctivitis — inflammation of the mucous membranes on the inner portion of the eyelids.
- Cyanosis — condition in which body surfaces turn blue because of lack of oxygen.
- Dyspnea — condition in which there is difficult or labored respiration.
- Edema — accumulation of fluids in the body cavities. Pulmonary edema is the accumulation of fluids in the lung.
- Fibrosis — condition in which fibrous tissue is formed as in scarring.
- Glottis — edema or buildup of fluids in the region of the larynx.
- Jaundice — yellowing of skin due to accumulation of bile pigments in the blood because of liver damage — the primary initial symptom of all the liver poisons discussed in the preceding unit.
- Methemoglobinaemia — anemia resulting from conversion of hemoglobin to methemoglobin as in carbon monoxide toxicity.
- Rale — an abnormal breathing sound as in severe asthma.

Respiratory Poison	Symptom
Upper Respiratory Poisons	
hydrogen chloride	At 35 ppm eye and throat irritation occurs. 100 ppm is tolerable for about an hour. Higher concentrations result in pulmonary edema.
hydrogen bromide	Generally similar to HCl but quantitative data are lacking.
hydrogen fluoride	Unlike HCl, HF cannot be tolerated even for short time periods. Produces severe skin burns and ulcers of the upper respiratory tract.
sulfur dioxide	Less than 1 ppm detectable by taste. 3 ppm detectable by smell. 5 to 10 ppm highly irritating to nasal mucous membranes and throat. Produces pulmonary edema and glottis. In actual practice, the SO_2 reacts with moist air to form sulfurous and sulfuric acid mists which penetrate into the lower respiratory tract. In this event, the symptoms are those of a lower respiratory poison.
nitrogen dioxide	Similar to SO_2 but quantitative data are lacking.
Lower Respiratory Poisons	
acrolein	Highly irritating to eyes, mucous membranes, and entire respiratory system. Quantitative data lacking.
chlorine	Odor detectable at 3 ppm. Immediate throat and nose irritation at 15 ppm. Hazardous at 50 ppm for even short exposures. High exposures result in pulmonary edema and audible chest rales.
bromine, fluorine	Similar to chlorine but quantitative data are lacking.
phosgene	Irritating to eyes at 3 ppm. Fatal at 50 ppm for even short exposures. High exposure results in pulmonary edema, dyspnea, and chest rales.
chloropicrin	Toxicity level between that of chlorine and phosgene.
hydrogen cyanide	Death is usually rapid and essentially symptomless. Mild exposure produces headache, dizziness, suffocation, and nausea.
methyl cyanide, acrylonitrile	Toxic agent is HCN, thus symptoms would be similar to those of HCN.
epichlorohydrin	Toxic agent is HCl, thus symptoms would be similar to those of HCl.
hydrogen sulfide	Highly toxic. Exposure to low concentrations produces conjunctivitis, vision problems, and digestive disturbances. Exposure to higher concentrations produces bronchitis and pulmonary edema.
methyl mercaptan, phosphorus pentasulfide	Toxic agent is H_2S thus symptoms would be similar to those of H_2S.
Affect Oxygen-Carrying Capacity of Blood	
aniline	Initial symptoms are due to conversion of hemoglobin to methemoglobin. Symptoms are methemoglobinaemia and anoxemia. Longer term exposure leads to jaundice and liver and bladder malfunctions.
benzene, toluene	Initial symptoms are dizziness, mental confusion, tightening of the leg muscles and then a stage of excitement. Continued exposure ultimately leads through a complex series of symptoms to coma.

Table 31-5. Clinical and Physiological Symptoms of Respiratory Poisons (continued)

Respiratory Poison	Symptom
carbon monoxide	Concentrations of 500 ppm for one hour are tolerable. Similar exposure to 1000 ppm is hazardous and 4000 ppm is likely to be fatal. Carbon monoxide is particularly hazardous because of the lack of distinctive symptoms.
hydrazine, phenylhydrazine	Symptoms are anemia and general weakness resulting from blood, liver, and kidney damage.
nitro compounds	Produce a wide variety of symptoms resulting from anemia, jaundice, and degeneration of liver, kidney, central nervous system. Since the hazardous properties of nitro compounds are largely those of explosive materials, the toxicity hazards have not been as extensively defined.

Dust and Particulate Matter

asbestos	early stage — shortness of breath and dry cough. middle stage — characteristic X-ray shadows, increasing shortness of breath, fibrosis of lung after several years. late stage — increasing fibrosis of lung, emphysema and related breathing difficulties such as dyspnea, cyanosis, and anoxemia.
alumina, coal, mica, soot	symptoms are not as well defined as for asbestos and silica but are generally similar.

Table 31-5. Clinical and Physiological Symptoms of Respiratory Poisons (continued)

REVIEW

1. The portion of the body which is most sensitive to a lack of oxygen is the
   - a. Lungs.
   - b. Blood.
   - c. Central nervous system.
   - d. Muscle.

2. Which of the following terms indicates a low level of oxygen in the blood?
   - a. Hyperoxia.
   - b. Hypoxia.
   - c. Anoxia.
   - d. Onoxia.

3. Internal respiration involves which of the following body structures?
   - a. Blood.
   - b. Lungs.
   - c. Trachea.
   - d. Bronchial tubes.

4. Respiratory poisons can function by attacking
   - a. Alveoli.
   - b. Blood.
   - c. Larynx.
   - d. All of these.

5. Sulfuric acid aerosol is a
   - a. Corrosive poison.
   - b. Respiratory poison.
   - c. Lower respiratory poison.
   - d. Any of these.

6. Sulfur dioxide is
   - a. SO_2.
   - b. An upper respiratory poison.
   - c. An irritant.
   - d. All of these.

7. The most hazardous respiratory poisons are those which are

    a. The least soluble in water.   c. Gases.
    b. The most soluble in water.   d. Liquids.

8. Phosgene is a particularly hazardous toxic combustion product because

    a. It is soluble in water.
    b. It is insoluble in water.
    c. It reacts with water to form HCl and CO.
    d. It does not react with water and therefore is not diluted by the water used to extinguish a fire.

9. Cyanides are

    a. Lower respiratory poisons.   c. Acids.
    b. Upper respiratory poisons.   d. All of these.

10. Barium cyanide produces toxic gaseous products when mixed with

    a. Water.   c. Bases.
    b. Acids.   d. Alcohols.

11. Potassium sulfide produces toxic gaseous products when mixed with

    a. Water.   c. Bases.
    b. Acids.   d. Alcohols.

12. Carbon monoxide is a hazardous respiratory poison because

    a. It reacts with water.
    b. It reacts with oxygen.
    c. It reacts with hemoglobin reversibly.
    d. It reacts with hemoglobin irreversibly.

13. Which of the following respiratory poisons results in anoxia?

    a. Carbon monoxide.   c. Phenylhydrazine.
    b. Aniline.   d. All of these.

14. Inhalation of which of the following dusts is potentially most hazardous?

    a. Beryllium oxide.   c. Asbestos.
    b. Silicon dioxide.   d. Mica.

15. Jaundice is an indication of

    a. Hemolysis.   c. Cyanosis.
    b. Liver damage.   d. Anoxemia.

16. Inhalation of excessive quantities of particulate matter is likely to lead to

    a. Hyperoxia.   c. Hemolysis.
    b. Glottis.   d. Fibrosis.

Unit 32 Nerve Poisons

Most physical and mental processes in the human body are controlled by the nervous system. The two primary functions of the nervous system are response to stimuli and conductivity of nerve impulses. High speed impulses continuously travel throughout the body keeping the various structures informed of external and internal changes. Many of the essential body functions controlled by the nervous system (heartbeat, breathing, and body temperature) are carried out below the level of consciousness by the autonomic part of the nervous system.

Nerve impulses travel through nerve cells called neurons. The velocity and method of propagation of the nerve impulse gives some indication of its fundamental nature. A nerve impulse is best defined as a self-propagating wave of negative charge traveling along a neuron (nerve cell) or across a synapse (nerve junction). This basic nerve structure is shown in figure 32-1. Blockage of essential nerve impulses results in rapid death.

Nerve poisons are particularly hazardous because they damage or destroy the control center for all of the essential body functions. The use of organic phosphorus compounds as nerve poisons began in World War II with the development of Sarin, Soman, and Tabum as potential chemical warfare agents. These compounds are usually called organic phosphorus compounds or organic phosphates because they can be viewed as derivatives of phosphoric acid (H_3PO_4). This terminology is used here although correct application of the rules of chemical nomenclature would demand that they be called phosphine oxides. It is important, however, to know that compounds of this type are phosphine oxides because most handbooks list them under that heading rather than as organic phosphates.

Some of the more common organic phosphate insecticides are listed in table 32-1. With the possible exception of Malathion, all are just as toxic as the materials developed as war gases. Insecticides are organic phosphates in the form of solids rather than gases because they can be applied to agricultural crops more effectively in solid form. Many other organic phosphate insecticides are marketed under a variety of trade names. Although toxicity levels vary considerably, all are hazardous nerve poisons and should be treated accordingly.

Many other materials are nerve poisons in the sense that they are central nervous system depressants or narcotics. They function in various ways to interfere with the transmission of nerve impulses at points other than the synapse. Some typical hazardous central nervous system depressants are listed in table 32-2. They tend to be considerably less hazardous than the synapse nerve poisons because their action is at less critical sites.

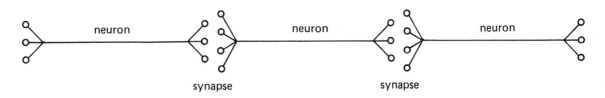

Fig. 32-1. Simplified Structure of a Nerve Neuron and Synapse.

Name	Application	Threshold Toxicity Limit	
		ppm for gases	mgm per cubic meter in air for solids
tetraethylpyrophosphate (TEPP)	insecticide	0.004	0.05
parathion	insecticide	- - -	0.1
malathion	insecticide	- - -	10.0
systox	insecticide	0.01	0.1
dimethyldichlorovinyl phosphate (DDVP)	insecticide	0.1	1

Table 32-1. Organic Phosphate Nerve Poisons.

Name	Formula	Application	Threshold Toxicity Limit	
			ppm for gases	mgm per cubic meter in air for solids
ethyl acetate	$CH_3COOCH_2CH_3$	solvent	400	1400
propyl acetate	$CH_3COOCH_2CH_2CH_3$	solvent	200	840
methyl acetate	CH_3COOCH_3	solvent	200	610
isopropyl alcohol	$(CH_3)_2CHOH$	solvent	400	980
acetone	CH_3CCH_3 $\overset{\|\|}{O}$	solvent	1000	2400
cyclohexane	(ring structure)	solvent	300	1050
carbon disulfide	CS_2	solvent, fumigant	20	60
formaldehyde	HCHO	sanitizer, chemical intermediate	2	3
acetaldehyde	CH_3CHO	chemical intermediate	100	180
furfural	(furan-CHO structure)	polymerization monomer	5	20
pyridine	(pyridine ring structure)	solvent, chemical intermediate	5	15

Table 32-2. Toxicity of Materials Which Are Primarily Central Nervous System Depressants.

REVIEW

1. Nerve poisons are particularly hazardous because

   a. All physical processes in the body are controlled by the nerves.
   b. All mental processes in the body are controlled by the nerves.
   c. The heartbeat is controlled by the nerves.
   d. All of the above.

2. The most hazardous nerve poisons are.

   a. Carbon monoxide and hydrogen cyanide.
   b. Phosgene.
   c. Organic phosphates.
   d. Hydrazine derivatives.

3. Nerve poisons are most likely to be encountered in handling

   a. Insecticides.   c. Electroplating chemicals.
   b. Plastics monomers.   d. Any of these.

4. The least toxic of the synapse nerve poisons is

   a. Sodium cyanide.   c. Parathion.
   b. Systox.   d. Malathion.

5. A central nervous system depressant such as carbon disulfide could be equally correctly called a

   a. Narcotic.   c. Respiratory poison.
   b. Heavy metal poison.   d. Liver poison.

6. The central nervous system depressant, ethyl acetate, is an

   a. Alcohol.   c. Aldehyde.
   b. Acid.   d. Ester.

7. Formaldehyde is most similar in hazardous toxic properties to

   a. Acetaldehyde.   c. Arsenic.
   b. Malathion.   d. Nitrobenzene.

Unit 33 Liver Poisons

The liver in the human body is an important organ involved in the purification and detoxification of the bloodstream. In the normal sequence of body digestive functions

- Food is digested in the stomach and small intestine.

- Nutrients (as well as potentially toxic substances) are absorbed into the bloodstream in the small intestine.

- Waste products are eliminated via the large intestine and kidney.

All chemicals absorbed into the body via the intestinal tract are processed by the liver to remove potentially toxic or harmful materials prior to going to the heart for general circulation as shown in figure 33-1. This purification or detoxification process is an important body defense mechanism. In the case of bacterial infections, the liver is an important means of removing bacterial toxins and dead bacterial cells. The toxins are converted by chemical reactions to a less toxic form in which excretion is facilitated. The dead bacterial cells are removed by simple filtration.

CHEMICAL COMPOUNDS

Bacterial toxins contain a wide range of chemical compounds such as amines, phenols, aromatic benzene rings and acids. As part of the evolutionary struggle for survival, the human body developed a mechanism for inactivating the chemical toxins until they could be excreted by the kidney. The same basic body detoxification mechanisms often serve as a defense against modern industrial chemicals which are often chemically similar. On the other hand, the body has no defense against some types of poisons such as hydrogen cyanide (HCN) and organic phosphates such as the insecticide parathion. In cases where the body has no defense mechanism against a chemical, even trace quantities of that substance can be lethal. Some of the typical body defense mechanisms against potentially toxic materials are summarized in table 33-1. Literally thousands of different chemical compounds could be listed in a similar manner. All are detoxified in a basically similar sequence of reactions.

- Reaction in liver.

- Conversion to a nontoxic compound.

- Transportation to kidney via the bloodstream.

- Excretion via kidney and urinary tract.

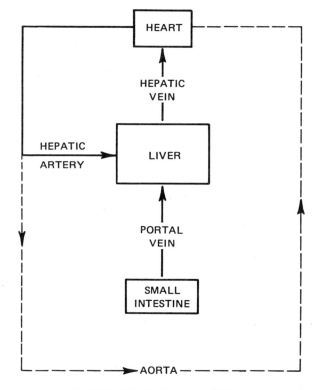

Fig. 33-1. Block Diagram of Liver Function In Human Body.

Toxin	Reacted in Liver With	Nontoxic Compound Formed	Excreted by
ammonia NH_3	carbon dioxide, water	urea H_2NC-NH_2 $\parallel$ O	kidney
organic acids such as benzoic acid ⬡COOH	glycine CH_2-COOH $\mid$ NH_2	hippuric acid O H $\parallel$ $\mid$ $C-N-CH_2COOH$ ⬡	kidney
phenols ⬡OH	cysteine CH_2SH $\mid$ $CHNH_2$ $\mid$ $COOH$	phenol sulfuric acid OSO_3H ⬡	kidney
halocarbons such as chlorobenzene ⬡Cl	cysteine CH_2SH $\mid$ $CHNH_2$ $\mid$ $COOH$	chlorophenyl mercapturic acid $S-CH_2$ O $\mid$ $\parallel$ ⬡CHNHCCH_3 Cl $COOH$	kidney
aromatic hydrocarbons such as naphthalene ⬡⬡	cysteine CH_2SH $\mid$ $CHNH_2$ $\mid$ $COOH$	naphthyl mercapturic acid $S-CH_2$ O $\mid$ $\parallel$ ⬡⬡CHNHCCH_3 $COOH$	kidney
aromatic amines such as pyridine ⬡N	methionine SCH_3 $\mid$ CH_2 $\mid$ CH_2 $\mid$ $CHNH_2$ $\mid$ $COOH$	N-methyl pyridine CH_3 $\mid$ N ⬡	kidney

Table 33-1. Typical Body Defense Mechanisms Against Toxic Materials.

HEAVY METALS

The liver also reacts with heavy metals such as lead, mercury, and cadmium in order to remove them from the bloodstream. Special proteins are present in the liver which have a great affinity for heavy metals. The toxic heavy metals are removed from the bloodstream as insoluble metal salts of the protein. The insoluble metal salts are removed from the bloodstream and the immediate toxicity hazard to the body is averted. However, the liver does not have a mechanism for eliminating the insoluble heavy metal protein compounds. For that reason, the heavy metals are cumulative poisons. They tend to accumulate in the liver and damage the liver's essential functions. Small amounts are excreted in the urine but because of the limited solubility, the elimination process is relatively ineffective. Some of the common heavy metal poisons and threshold toxicity limits are given in table 33-2. The heavy metals are hazardous as liver poisons when they enter the body as soluble compounds. The same metals are hazardous as

Metal	Commonly Encountered Form	D.O.T. Toxicity Classification	ppm for gas	mgm per cubic meter of air for solids
antimony	insoluble forms: metal, oxides soluble forms: chloride gas: stibine, SbH_3	- - -	0.1	0.5 0.5
arsenic	insoluble forms: metal, oxides soluble forms: chloride, arsenous acid, some forms of sheep dip gas: arsine, AsH_3	Class B	0.05	25 0.2
barium	insoluble forms: metal, sulfate oxide soluble forms: chloride, sulfide carbonate	- - -		- - - 0.05
beryllium	insoluble forms: metal, oxide soluble forms: chloride, carbonate, hydride	Class B		0.002
cadmium	insoluble forms: metal, oxide soluble forms: chloride, cyanide, nitrate	- - -		0.05
chromium	insoluble forms: metal is nontoxic soluble forms: chromates, oxides, sulfates, carbonyl	- - -		0.1
lead	insoluble forms: metal, oxide soluble forms: chloride, carbonate gas: tetraethyl lead	- - -	0.1	0.15
mercury	insoluble forms: metal oxide soluble form: nitrate	Class B		0.05
thallium	soluble form: oxide, sulfate nitrate, rat poison	Class B		0.1

Table 33-2. Toxicity of Heavy Metals.

respiratory poisons when they are inhaled as insoluble metal or metal oxide dust. Thus tetraethyl lead is primarily a liver poison because it is soluble; metallic lead tends to be toxic only when inhaled as a dust and is primarily a respiratory poison in that role. Precise data are not generally available for heavy metal toxicity in either the soluble or insoluble form. In that sense, the distinction is largely academic. To the extent that data are available, listings for both the soluble and insoluble forms are included in table 33-2.

ORGANIC CHLORINATED COMPOUNDS

The other major class of liver poisons contains the organic chlorinated compounds such as chlorinated solvents, insecticides, moth crystals, fumigants, and other specialty products. One of the major functions of the liver is the storage of fats. Fat soluble organic chlorine compounds are concentrated in the liver because of mutual solubility characteristics. With the exception of a few aromatic chlorine compounds, the human body does not have an effective detoxification mechan-

Name	Formula	Threshold Toxicity Limit	
		ppm for gas	mgm per cubic meter of air for solids
Aliphatic Compounds			
methyl chloride	CH_3Cl	100	210
methylene dichloride	CH_2Cl_2	100	360
carbon tetrachloride	CCl_4	10	65
ethyl chloride	CH_3CH_2Cl	1000	2600
chloroform	$CHCl_3$	25	120
perchlorethylene	$CCl_2{=}CCl_2$	100	670
dichloroethylene	$\begin{matrix} CH{=}CH \\ \mid \quad \mid \\ Cl \quad Cl \end{matrix}$	200	790
ethylene dichloride	$\begin{matrix} CH_2{-}CH_2 \\ \mid \quad \mid \\ Cl \quad Cl \end{matrix}$	50	200
ethylene dibromide	$\begin{matrix} CH_2{-}CH_2 \\ \mid \quad \mid \\ Br \quad Br \end{matrix}$	25	145
Aromatic Compounds			
chlorobenzene	Cl⬡	75	350
p-dichlorobenzene	Cl⬡Cl	75	450
o-dichlorobenzene	Cl⬡Cl	50	300

Table 33-3. Toxicity of Chlorinated Hydrocarbon Compounds.

ism. Since the primary concentration site is in the fatty portion of the liver, these compounds are hazardous toxic materials because they are severely damaging to the liver tissues.

Some of the common chlorinated hydrocarbons and toxicity data are given in table 33-3. The halogenated hydrocarbons are unique in that they are unusually inert chemicals. They do not enter into the body metabolism because of reactivity. Similarly, they are inert to the body defense mechanisms for the same reason. If it were not for the special property of selective collection in the fatty tissues of the liver, toxicity would not be such a serious problem.

In the recent past, insecticides such as DDT, which are highly chlorinated hydrocarbons, were widely used in the agriculture industry. Insecticide activity depended on poisoning the livers of undesirable insects. Chlorinated insecticides such as DDT are highly damaging to the biological environment because they are not broken down into harmless substances by the soil microorganisms. The threshold toxicity limits of some of the more common insecticides, fumigants, and other industrially important halogen compounds are given in table 33-4.

The substitution of a chlorine atom for a hydrogen atom in a hydrocarbon compound

Name	Formula	Application	Threshold Toxicity Limit	
			ppm for gases	mgm per cubic meter in air for solids
acetylene tetrachloride	$\underset{\displaystyle Cl_2 \quad Cl_2}{CH-CH}$	solvent	5	35
acetylene tetrabromide	$\underset{\displaystyle Br_2 \quad Br_2}{CH-CH}$	solvent	1	14
aldrin	$C_{12}H_8Cl_6$	insecticide		0.25
allyl chloride	$CH_2=CHCH_2Cl$	polymerization monomer	1	3
lindane (benzene hexachloride)	$C_6H_6Cl_6$	insecticide		0.5
benzyl chloride	CH_2Cl (benzene ring)	chemical intermediate	1	5
chlorodane	$C_{10}H_6Cl_8$	insecticide		0.5
chlorodiphenyl	Cl (biphenyl) Cl	chemical intermediate		0.5
chloroprene	$\underset{\displaystyle Cl}{CH_2=CH-C=CH_2}$	polymerization monomer	25	90
DDD	$(ClC_6H_4)_2CHCHCl_2$	insecticide		1.0
DDT	$(ClC_6H_4)_2CHCCl_3$	insecticide		1.0
dichlorodifluoromethane	CCl_2F_2	Freon-12	1000	4950
dieldrin	$C_{12}H_{10}OCl_6$	insecticide		0.25
fluorotrichloromethane	CCl_3F	Freon-11	1000	5600
dichlorofluoromethane	$CHCl_2F$	Freon-21	1000	4200
1, 2-dichloro -1, 1, 2, 2, -tetrafluoroethane	$CClF_2-CClF_2$	Freon-114	1000	7000
vinylchloride	$CH_2=CHCl$	polymerization monomer	pending*	pending*
1, 2-dinitrobenzene	NO_2 (benzene ring) NO_2	explosive		1

*assume 0, pending official assignment of TLV.

Table 33-4. Toxicity of Industrially Important Halogenated Compounds Which Are Primarily Liver Poisons.

increases the narcotic and anesthetic action of the parent compound. The primary toxic effect of hydrocarbons is narcotic and anesthetic action. Chlorine and other halogen substitutions on the molecule have the effect of enhancing this activity. In addition, the halogen does separate damage as a systemic liver poison. It is difficult and sometimes dangerous to try to predict the relative toxicity of various types of aliphatic and aromatic halogen compounds but some generalities are possible.

- Unsaturated chlorine compounds are generally more narcotic and less toxic as liver poisons than the corresponding saturated compounds. Thus perchlorethylene may be predicted to be more narcotic but less of a systemic poison than acetylene tetrachloride.

- Within a given series of saturated aliphatic compounds the narcotic effect increases as the number of chlorine atoms increases. It is not possible to make a similar prediction about the systemic poison level on the basis of increasing chlorine content, however. Thus methylene dichloride (CH_2Cl_2) is not necessarily a worse systemic poison than methyl chloride (CH_3Cl).

- The toxicity of aromatic chlorine compounds is difficult to predict, but, in general, they are no worse than the corresponding aromatic hydrocarbon. This similarity in toxicity is a direct result of the body detoxification system which has a means of detoxifying aromatic compounds. The toxicity of benzene may be predicted to be about the same as of chlorobenzene. The body gets rid of the toxic benzene ring and as a secondary benefit gets rid of the chlorine atom at the same time.

- Organic fluorine compounds are generally less toxic than the other halohydrocarbons. The chlorine, bromine, and iodine substituted compounds are similar in their toxicity level of corresponding compounds. Thus CH_3F may be predicted to be much less toxic than CH_3Cl, CH_3Br, or CH_3I.

- All chlorinated insecticides and similar compounds are among the most toxic chlorinated hydrocarbons. They were selected from among available compounds on the basis of high toxicity.

TOXICITY LIMITS

In the United States, the threshold toxicity limits set by the American Conference of Governmental Industrial Hygienists are the most universally accepted quantitative toxicity criteria. These toxicity values are usually referred to as *ACGIH accepted values*. All threshold limits (TL) listed in this textbook are ACGIH accepted TL values and as such are typical of the best values available at this time. These data are revised annually, so for critical applications the reader is strongly advised to refer to the most recent official publication of the American Conference of Governmental Industrial Hygienists.

For toxic gases, the TL values are given in parts per million (ppm) of the toxic material per million parts of air. Thus 1 ppm is 1 part in 1,000,000 parts of air.

For finely divided dusts, particulate matter, and aerosols the TL value is expressed in milligrams of solid per cubic meter of air. This unit of measurement is used because it is more feasible to weigh solids suspended in air than to try to measure the volume.

For some dusts such as silicon dioxide and asbestos, the TL is expressed in millions of particles per cubic foot of air. This convention is based on common usage rather than scientific necessity or logic.

Another widely used criteria for toxicity rating is that of the United States Department of Transportation (DOT). (The official DOT and other related safety criteria are discussed in detail in unit 37.) This is a qualitative rating system which clearly identifies the relative hazard and type of hazard associated with various types of materials. Hazards are identified by a simplified label system.

- Class A — extremely dangerous poisons.

- Class B — less dangerous poisons.

- Corrosive liquids – no class identification because in the sense of strict definition they are not poisons.

- Class C – irritants and tear gases.

- Class D – radioactive materials.

REVIEW

1. Toxic waste products are removed from the bloodstream by the

   a. Liver.
   b. Kidney.

   c. Lungs.
   d. Small intestine.

2. When a toxic substance is ingested, the point of entry into the bloodstream is via the

   a. Liver.
   b. Kidney.

   c. Skin.
   d. Small intestine.

3. The liver removes toxic materials from the blood

   a. Continuously.
   b. Intermittently.
   c. At night.
   d. Only after exposure to unusually high levels of hazardous toxic materials.

4. The liver has no defense against

   a. Bacterial toxins.
   b. Carbon monoxide.
   c. Hydrogen cyanide.
   d. All of the above.

5. The liver provides a protective function by removing toxic heavy metals from the bloodstream. These heavy metals are liver poisons because
   a. They are so insoluble that the liver cannot eliminate them efficiently.
   b. The liver tissues are more susceptible to heavy metal poisons than the rest of the body.
   c. They are so soluble that they immediately go back into solution in the bloodstream.
   d. They are similar to bacterial toxins.

6. An example of a heavy metal poison is

   a. Iron.
   b. Chromium.

   c. Sodium.
   d. Calcium.

7. When a heavy metal powder such as beryllium metal is inhaled

   a. It is primarily a liver poison.
   b. It is absorbed into the bloodstream via the small intestine.
   c. It is primarily a respiratory poison.
   d. It is primarily an irritant.

8. Light elements are

   a. Sometimes toxic hazardous materials.
   b. Never toxic hazardous materials.
   c. Always toxic hazardous materials.
   d. Essential to proper body function.

9. When a toxic nonmetal such as phosphorus is ingested it is

   a. Detoxified by the liver.
   b. Stored in the blood.
   c. Stored in the small intestine walls.
   d. Exhaled as phosphorus dioxide through the lungs along with carbon dioxide.

10. Chlorinated hydrocarbons are liver poisons because

    a. They are fats.
    b. They are soluble in fats.
    c. They are insoluble in fats.
    d. They are often used as insecticides.

11. Chlorinated hydrocarbons are likely to be encountered in handling

    a. Insecticides.   c. Degreasers.
    b. Solvents.   d. Any of these.

Unit 34 Corrosive Poisons

The corrosive poisons include a wide variety of materials which are hazardous because they are chemically reactive with body tissues. They can react directly, or the secondary product formed after reaction with water can be the reactive agent. Some corrosive poisons which are hazardous primarily because of their action on the respiratory system have already been discussed. Some of the common corrosive poisons are listed in table 34-1.

Name	Formula	Threshold Toxicity Limit	
		ppm for gases	mgm per cubic meter in air for solids
Compounds Related to Sulfuric Acid			
sulfur dioxide	SO_2	5	13
sulfur trioxide	SO_3	- - -	- - -
sulfurous acid	H_2SO_3	- - -	- - -
sulfuric acid	H_2SO_4	- - -	1
dimethyl sulfate	$(CH_3)_2SO_4$	pending*	pending*
Compounds Related to Nitric Acid			
nitric oxide	NO	25	30
nitrogen dioxide	NO_2	5	9
nitrogen tetroxide	N_2O_4	5	9
nitric acid	HNO_3	2	5
nitrous acid	HNO_2	- - -	- - -
Compounds Related to Acetic Acid			
acetic acid	CH_3COOH	10	25
acetic anhydride	$(CH_3CO)_2O$	5	20
Compounds Related to Phosphoric Acid			
phosphoric acid	H_3PO_4		1
phosphorus pentachloride	PCl_5		1
phosphorus trichloride	PCl_3	0.5	3
Phenolic Compounds			
phenol	OH⬡	5	19
cresol	CH_3⬡OH and CH_3⬡OH	5	22
Alkaline Compounds			
ammonia	NH_3	25	18
diethylamine	$(CH_3CH_2)_2NH$	25	75
isopropyl amine	CH_3\>CH_2NH_2 CH_3/	5	12
ethanolamine	$HOCH_2CH_2NH_2$	3	6
sodium hydroxide	$NaOH$	- - -	- - -
Miscellaneous			
hydrogen peroxide	H_2O_2	1	1.4
perchloryl fluoride	ClO_3F	3	14

*assume 0 pending official assignment of TLV.

Table 34-1. Toxicity of Materials Which Are Primarily Corrosives.

SULFURIC ACID

The corrosive action of sulfuric acid in contact with body tissues is well known. Severe burns resulting in extensive tissue damage result. In addition to damage to the respiratory system discussed previously, consideration must also be given to potential damage to other tissues such as the eyes and skin. Although quantitative data are not available for sulfurous acid, the magnitude of corrosive hazard is similar to that of sulfuric acid. The sulfur oxides can also be considered to be corrosive poisons because they react with water to form the corresponding acids.

The organic derivatives of sulfuric acid are also similar in corrosive toxicity. Some examples are benzene sulfonic acid, benzene sulfonyl chloride, and dimethyl sulfate.

NITRIC ACID

The hazardous property relationships of the nitric acid series of compounds are quite similar to those of sulfuric acid. Both nitric and nitrous acids are highly corrosive toxic materials. The oxides of nitrogen react with water to form the corresponding acids in the same manner as any other nonmetal oxide.

The organic derivatives of nitric acid, nitrobenzene and methyl nitrate, are not corrosive poisons as was the case with sulfuric acid. These derivatives are no longer acids;

they are the nitro- and nitric ester derivatives which tend to be liver and blood poisons.

ACETIC ACID

Acetic acid and the anhydride, acetic anhydride, are corrosive poisons. Acetic anhydride is particularly corrosive because it reacts with body water to form acetic acid.

PHOSPHORIC ACID

Phosphoric acid is a very hazardous corrosive toxic material. The organic phosphate derivatives discussed under the heading *Nerve Poisons* are among the most toxic hazardous materials. Other common phosphorus derivatives which are corrosive toxic materials are phosphorus trichloride, phosphorus pentachloride, phosphorus pentoxide, and phosphorus oxychloride.

AMMONIA

Another major class of corrosive toxic materials includes ammonia and the organic derivatives of ammonia which are known as amines. Most of the amines tend to be corrosive poisons in addition to other adverse effects such as blocking the oxygen-carrying capacity of the blood. As a class, the amines are just as corrosive and hazardous as the metal hydroxides which are more familiar materials.

REVIEW

1. All corrosive poisons are similar in the sense that they

   a. Are acids.
   b. Chemically react with body tissues.
   c. Are not respiratory poisons.
   d. Are not liver poisons.

2. An example of a corrosive poison is

   a. Sulfuric acid.
   b. Hydrochloric acid.
   c. Nitric acid.
   d. All of these.

3. Which of the following sulfate compounds is most toxic?

   a. Dimethyl sulfate.   c. Calcium sulfate.
   b. Sodium sulfate.   d. Iron sulfate.

4. Which of the following is a corrosive toxic hazardous material?

   a. Nitromethane.   c. Nitric acid.
   b. Nitrobenzene.   d. All of these.

5. Glacial acetic acid is a greater corrosive toxic hazard than vinegar because of

   a. Lower water content.
   b. Higher vapor pressure.
   c. Higher flammability.
   d. Yields toxic combustion products.

6. Which of the following compounds yields phosphoric acid in the presence of water?

   a. Phosphorus pentachloride.   c. Phosphorus pentoxide.
   b. Phosphorus trifluoride.   d. All of these.

7. An example of an alkaline corrosive toxic material is

   a. Sodium hydroxide.   c. Methylamine.
   b. Ammonia.   d. All of these.

8. Which of the following is a corrosive toxic material but is not an acid or an alkaline material?

   a. Hydrogen chloride.   c. Ethanolamine.
   b. Hydrogen peroxide.   d. Sodium carbonate.

SECTION 9 RADIOACTIVE MATERIALS

Unit 35 Principles of Radioactivity

The structure of the elements has been presented as a means of establishing the fundamental basis of all chemical reactions. In figure 1-1 and related explanatory material, it is pointed out that all elements are composed of

- A nucleus consisting of positive particles called protons and neutral particles called neutrons.

- Valence electrons equal in number to the number of protons in the nucleus.

All of the ordinary chemical reactions of the type discussed to this point have been governed by the valence electron configuration of an atom. For that reason, it has not been necessary to pay much attention to the nucleus of the atom. Reactions such as sodium with chlorine are determined by the valence electrons.

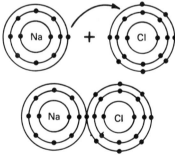

In contrast, radioactivity is determined entirely by the neutron and proton configuration of the nucleus. The principles of radioactivity must be explained in terms of nuclear chemistry.

The simplest starting point is with a study of the nuclear reactions of the proton. It may be recalled that a proton is the positively charged nucleus of a hydrogen atom.

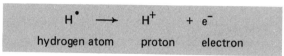

$$H^{\bullet} \longrightarrow H^{+} + e^{-}$$
hydrogen atom   proton   electron

In nuclear chemistry, the proton is represented as H_1^1.

$$H \quad \frac{\text{number of protons + neutrons}}{\text{number of protons}}$$

In unit 1, it was indicated that hydrogen contains one proton and no neutrons. Thus, the nuclear configuration of hydrogen is H_1^1. The corresponding nuclear configuration of the other elements are given in the equivalent nuclear chemistry representation in table 35-1.

The lower subscript indicates the number of protons in the nucleus and is the atomic number of the element. The atomic number of the nucleus uniquely determines the particular element. The number of neutrons determines the particular isotope of the element. Thus, hydrogen, deuterium, and tritium are all isotopes of hydrogen.

hydrogen	H_1^1
deuterium	H_1^2
tritium	H_1^3

Similarly, O_8^{16}, O_8^{15}, and O_8^{17} are isotopes of oxygen.

As can be seen from table 35-1, most of the elements exist in more than one stable isotopic form. This is the basic reason for the fact that the atomic weights of the elements are not integers. Chlorine is an excellent example. Chlorine exists in nature as a mixture of its stable isotopes in the proportions shown on page 191.

Element	Symbol	Atomic Number	Stable Isotopes	Element	Symbol	Atomic Number	Stable Isotopes
hydrogen	H	1	1, 2	antimony	Sb	51	121, 123
helium	He	2	3, 4	tellurium	Te	52	120, 122, 123, 124,
lithium	Li	3	6, 7				125, 126, 128, 130
beryllium	Be	4	9	iodine	I	53	127
boron	B	5	10, 11	xenon	Xe	54	124, 126, 128, 129,
carbon	C	6	12, 13				130, 131, 132, 134,
nitrogen	N	7	14, 15				136
oxygen	O	8	16, 17, 18	cesium	Cs	55	133
fluorine	F	9	19	barium	Ba	56	130, 132, 134, 135,
neon	Ne	10	20, 21, 22				136, 137, 138
sodium	Na	11	23	lanthanum	La	57	139
magnesium	Mg	12	24, 25, 26	cerium	Ce	58	136, 138, 140, 142
aluminum	Al	13	27	praseodymium	Pr	59	141
silicon	Si	14	28, 29, 30	neodymium	Nd	60	142, 143, 144, 145,
phosphorus	P	15	31				146, 148, 150
sulfur	S	16	32, 33, 34, 36	promethium	Pm	61	Radioactive
chlorine	Cl	17	35, 37	samarium	Sm	62	144, 147, 148, 149,
argon	Ar	18	36, 38, 40				150, 152, 154
potassium	K	19	39, 40, 41	europium	Eu	63	151, 153
calcium	Ca	20	40, 42, 43, 46, 48	gadolinium	Gd	64	152, 154, 155, 156,
scandium	Sc	21	45				157, 158, 160
titanium	Ti	22	46, 47, 48, 49, 50	terbium	Tb	65	159
vanadium	V	23	51	dysprosium	Dy	66	158, 160, 161, 162,
chromium	Cr	24	50, 52, 53, 54				163, 164
manganese	Mm	25	55	holium	Ho	67	165
iron	Fe	26	54, 56, 57, 58	erbium	Er	68	162, 164, 166, 167,
cobalt	Co	27	57, 59				168, 170
nickel	Ni	28	58, 60, 61, 62, 64	thulium	Tm	69	169
copper	Cu	29	63, 65	ytterbium	Yb	70	168, 170, 171, 172,
zinc	Zn	30	64, 66, 67, 68, 70				173, 174, 176
gallium	Ga	31	69, 71	lutetium	Lu	71	175, 176
germanium	Ge	32	70, 72, 73, 74, 76	hafnium	Hf	72	174, 176, 177, 178,
arsenic	As	33	75				179, 180
selenium	Se	34	74, 76, 77, 78, 80, 82	tantalum	Ta	73	181
bromine	Br	35	79, 81	tungsten	W	74	180, 182, 183, 184, 186
krypton	Kr	36	78, 80, 82, 83, 84, 86	rhenium	Re	75	185, 187
rubidium	Rb	37	85, 87	osmium	Os	76	184, 186, 187, 188,
strontium	Sr	38	84, 86, 87, 88				189, 190, 192
yttrium	Y	39	89	iridium	Ir	77	191, 193
zirconium	Zr	40	90, 91, 92, 94, 96	platinum	Pt	78	192, 194, 195, 196,
niobium	Nb	41	93				197, 198
molybdenum	Mo	42	92, 94, 95, 96, 97, 98, 100	gold	Au	79	197
technetium	Tc	43	Radioactive	mercury	Hg	80	196, 198, 199, 200, 201, 202, 204
ruthenium	Ru	44	95, 98, 99, 100, 101, 102	thallium	Tl	81	203, 205
				lead	Pb	82	204, 206, 207, 208
rhodium	Rh	45	103	bismuth	Bi	83	209
palladium	Pd	46	102, 104, 105, 106, 108, 110	polonium	Po	84	Radioactive
				astatine	At	85	Radioactive
silver	Ag	47	107, 109	radon	Rn	86	Radioactive
cadmium	Cd	48	106, 108, 110, 111, 112, 113, 114, 116,	francium	Fr	87	Radioactive
				radium	Ra	88	Radioactive
indium	In	49	113, 115	actinium	Ac	89	Radioactive
tin	Sn	50	112, 114, 115, 116, 117, 118, 119, 120, 122, 124	thorium	Th	90	232
				protactinium	Pa	91	Radioactive
				uranium	U	92	234, 235, 238

Table 35-1. Stable Isotopes of the Elements.

$$75\% \;-\; Cl_{17}^{35}$$
$$25\% \;-\; Cl_{17}^{37}$$

Isotopes of the elements are formed by the reaction or fusion of the nucleus of chemical elements. A major class of common nuclear reactions is the fusion of a proton with another atomic nucleus as shown in table 35-2. Nuclear reactions tend to be confusing to the student. For example, beryllium is converted to three different elements with the emission of three different elementary particles. The mechanism of the nuclear reactions is not well understood. The most probable explanation is that the proton is taken up by the beryllium nucleus to form an unstable complex which disintegrates almost instantaneously by one of the three routes listed in table 35-2.

Production	Disintegration
$H_1^1 + Be_4^9 \rightarrow Li_3^6 + He_2^4$	none, Li_3^6 is a stable isotope
$H_1^1 + Ar_{18}^{40} \rightarrow K_{19}^{40} + \eta_0^1$	none, K_{19}^{40} is a stable isotope
$H_1^1 + Be_4^9 \rightarrow B_5^{10} + \gamma$	$B_5^{10} \rightarrow C_6^{10} + e_{-1}^0$
$H_1^1 + Be_4^9 \rightarrow Be_4^8 + H_1^2$	$Be_4^8 \rightarrow He_2^4 + He_2^4$
$H_1^1 + Li_3^7 \rightarrow He_2^4 + He_2^4$	none, He_2^4 is a stable isotope

Table 35-2. Formation of Isotopes of the Elements by Proton Bombardment.

$$[complex_5^{10}] \begin{cases} \rightarrow Be_4^8 + H_1^2 \\ \rightarrow B_5^{10} + \gamma \text{ radiation} \\ \rightarrow Li_3^6 + He_2^4 \end{cases}$$

The isotopes formed can be either stable or unstable. The rules for predicting whether the product of a nuclear reaction is a stable or an unstable isotope are beyond the scope of this text and are not essential to an understanding of hazardous materials. The produc-

Production	Disintegration
neutron fusion reactions	
$\eta_0^1 + Co_{27}^{59} \rightarrow Co_{27}^{60} + \gamma$ Radiation	$Co_{27}^{60} \rightarrow N_{28}^{60} + e_{-1}^0$
$\eta_0^1 + Mg_{12}^{24} \rightarrow Na_{11}^{24} + H_1^1$	$Na_{11}^{24} \rightarrow Mg_{12}^{24} + e_{-1}^0$
deuteron fusion reactions	
$H_1^2 + C_6^{12} \rightarrow N_7^{13} + \eta_0^1$	$N_7^{13} \rightarrow C_6^{13} + e_{+1}^0$
$H_1^2 + Pb_{82}^{207} \rightarrow Pb_{82}^{208} + H_1^1$	none, Pb_{82}^{208} is a stable isotope
$H_1^2 + H_1^2 \rightarrow H_1^3 + H_1^1$	$H_1^3 \rightarrow He_2^3 + e_{-1}^0$
α-particle fusion reactions	
$He_2^4 + Mg_{12}^{25} \rightarrow Al_{13}^{28} + H_1^1$	$Al_{13}^{28} \rightarrow Si_{14}^{28} + e_{-1}^0$
$He_2^4 + Al_{13}^{27} \rightarrow P_{15}^{30} + \eta_0^1$	$P_{15}^{30} \rightarrow Si_{14}^{30} + e_{+1}^0$
$He_2^4 + N_7^{14} \rightarrow F_9^{17} + \eta_0^1$	$F_9^{17} \rightarrow O_8^{17} + e_{+1}^0$
γ-ray fusion reactions	
$\gamma_0^0 + H_1^2 \rightarrow H_1^1 + \eta_0^1$	none, H_1^1 is a stable isotope
$\gamma_0^0 + Be_4^9 \rightarrow Be_4^8 + \eta_0^1$	$Be_4^8 \rightarrow He_2^4 + He_2^4$
$\gamma_0^0 + P_{15}^{31} \rightarrow P_{15}^{30} + \eta_0^1$	$P_{15}^{30} \rightarrow Si_{14}^{30} + e_{+1}^0$
electron fusion reactions	
$e_{-1}^0 + Cu_{29}^{64} \rightarrow Ni_{28}^{64}$	none, Ni_{28}^{64} is a stable isotope

Table 35-3. Formation of Isotopes of the Elements by Typical Nuclear Reactions.

tion reactions listed in table 35-2 and the following tables are always conducted in a nuclear reactor, cyclotron, atom bomb, or nuclear bomb and as such are outside the realm of ordinary hazardous materials. This phase of the discussion is provided as a means of explaining the principles of radioactivity.

On the other hand, the disintegration reactions of unstable isotopes are the basis for the hazardous properties of radioactive materials. Stated in another way – the hazardous properties of radioactive materials result from the delayed disintegration of the unstable isotope to a stable isotope. The unstable isotopes of the elements are commonly called radioactive isotopes because they decay to stable isotopes with the simultaneous release of hazardous by-products.

The hazardous by-products of the radioactive disintegration of unstable isotopes to stable isotopes are listed.

- γ radiation, also called X-rays, and gamma rays.
- Protons, H_1^1, hydrogen nuclei.
- Neutrons, N_0^1.
- Deuterons, H_1^2, deuterium nuclei.
- α particles, He_2^4, helium nuclei.
- β rays, beta rays, high energy electrons.
- positron, e_{+1}^0, e^+.

Unstable isotopes are formed by reaction or fusion of the several nuclei listed above with another nucleus just as with the proton reactions shown in table 35-2. All of the reactions are basically similar in the sense that the reaction product is either a stable or an unstable isotope. Several typical reactions which lead to the formation of isotopes are summarized in table 35-3. It should also be noted that the products of the disintegration of the radioactive isotopes are the same hazardous elementary particles discussed previously (neutrons, proton, deuterons, α-particles, γ-rays, electrons, and positrons). The hazardous properties of these radioactive disintegration products determines the hazardous radioactive properties of any isotope.

REVIEW

1. The nucleus of a radioactive element consists of

   a. Protons and electrons.
   b. Neutrons and protons.
   c. Neutrons, protons, and electrons.
   d. Atoms and radioactive particles.

2. A neutron is composed of

   a. An electron and a proton.
   b. A proton and a gamma ray.
   c. An electron and an alpha particle.
   d. None of the above.

3. The symbol for gamma is

   a. α.
   b. β.
   c. g
   d. γ.

4. The symbol for a proton is

   a. p.
   b. H^+.
   c. H_1^1.
   d. Any of these.

5. The symbol for an alpha particle is

    a. He_2^4.
    c. H_1^2.
    b. α_0^0.
    d. e_0^+.

6. A radioactive isotope differs from a stable isotope of the same element because it contains a different number of

    a. Protons and neutrons.
    c. Protons.
    b. Positrons.
    d. Neutrons.

7. If an isotope has the symbol X_{16}^{18}, how many neutrons does it contain?

    a. 2.
    c. 18.
    b. 16.
    d. 34.

8. If an isotope has the symbol X_6^{12}, how many electrons does the isotope atom contain?

    a. 1.
    c. 6.
    b. 2.
    d. 12.

9. Which of the following isotopes has the greatest mass?

    a. H_1^1.
    b. H_1^2.
    c. H_1^3.
    d. All are the same because they are all hydrogen isotopes.

10. Radioactive isotopes are hazardous materials because

    a. They decay to form stable isotopes.
    b. They decay to form isotopes which can be either stable or unstable isotopes.
    c. They decay to form toxic elements.
    d. They decay to form flammable elements.

11. If an isotope has the symbol X_6^{12}, what chemical element is it?

    a. Carbon.
    c. Oxygen.
    b. Hydrogen.
    d. Nitrogen.

12. Radioactive isotopes decay to more stable states by the emission of

    a. Electrons.
    c. Gamma rays.
    b. Protons.
    d. Any of these.

Unit 36 Hazards of Radioactivity

The rapid development of radiation technology has led to extensive application of radioactive isotopes in medicine, industry, and warfare. Similarly, X-rays are widely used in both medicine and industry. Hazardous radiation from radioactive materials and X-ray sources is invisible and the effects are not immediately apparent in most cases. The situation is very similar to that of toxic hazardous materials. The health hazard and nature of the physiological response depend on several factors.

- Exposure time.
- Nature of radiation.
- Intensity of radiation.

In order to handle radioactive materials with safety it is necessary to understand the nature of radiation, how it can be measured, and the effects of exposure to various radiation levels.

The hazardous properties of radioactive materials are usually thought of in terms of nuclear radiation. All types of radiation share the common property of being absorbed and transferring energy to the absorbing body. Low energy radiation is easily absorbed, thus infrared light radiation from a heat lamp is largely absorbed and the absorbing body is heated. Since the infrared light rays are low energy, they are not particularly damaging to the body under normal exposures. On the other hand, massive exposure to low energy infrared radiation can be hazardous.

Ultraviolet light radiation is higher energy than infrared. Since it is higher energy, it penetrates deeper into the body and causes greater damage. Familiar examples are sunburn and blindness caused by looking directly into the sun. The damage caused by ultraviolet light rays results from

- Greater body penetration because of higher energy.

- Greater energy transfer because of greater body penetration coupled with higher energy content of each light photon.

The three main types of radiation that fire fighters may encounter are alpha, beta, and gamma. All three are different and each has different penetrating power. Alpha particles are the largest of the three and travel the smallest distance. They have the least penetration power of all; they will not even travel through a piece of paper. Beta particles have greater penetration power than alpha but very rarely penetrate the protective gear of fire fighters. Gamma rays are different. They penetrate turnout gear easily. They are a very high and dangerous source of energy.

GAMMA RAYS

Gamma rays (γ-rays or X-rays as they are often called are the next higher energy level to light rays. In the medical X-ray application, the gamma rays can be seen to have sufficiently high energy to penetrate the human body. The X-rays which are absorbed in the body transfer energy to the absorbing body in the same manner as ultraviolet rays. The average dental X-ray delivers 5 roentgens, whereas cosmic rays give only 0.1 roentgens per year. Medical X-rays are not entirely without hazard. However, the hazard is minimal because the exposure intensity is kept as low as possible. Gamma rays are more hazardous than ultraviolet rays because of their higher energy content. They cause burns just as ultraviolet rays do, and burns caused by the gamma rays can extend throughout the body.

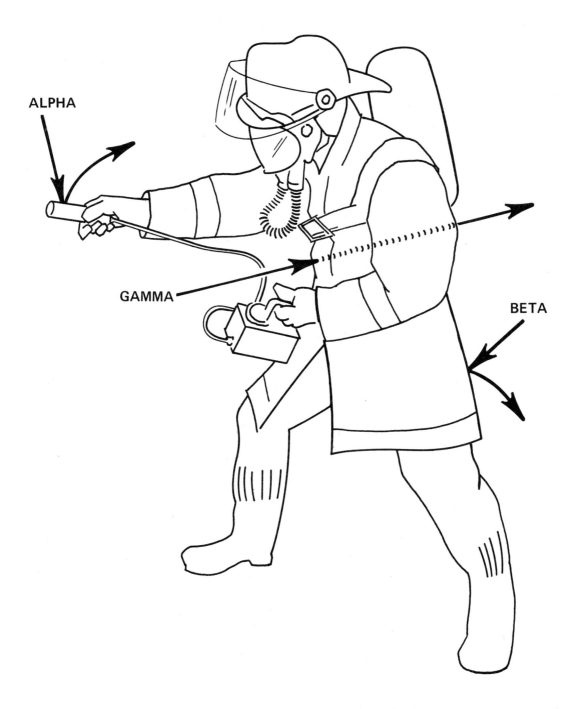

Fig. 36-1. Penetration Power of Alpha, Beta, and Gamma Radiation.

Gamma rays travel at the speed of light (186,000 miles per second).

The energy of radiation is expressed by physicists in units of electron-volts. An electron-volt is defined as the energy acquired by an electron when it is accelerated by an electrical potential of one volt. This is a very small unit. Radioactive materials emit radiations which fall in the million electron-volt range.

> electron-volt, eV
>
> million electron-volts, MeV

Gamma ray radiation energies from radioisotopes can be as high as 3 MeV. Some are emitted at relatively low energies and are correspondingly less hazardous X-ray radiation sources.

ALPHA PARTICLES

Alpha particles (He_2^4), protons (H_1^1), and deuterons (D_1^2 or H_1^2) are emitted by many of the higher atomic number radioactive isotopes. They are so small that a hundred billion billion alpha particles would be no larger than the head of a pin.

In contrast to light and gamma rays which have no mass and electrical charge, the particles just mentioned have a relatively large mass and a positive electrical charge, compared to the negatively charged electron.

> α — alpha particles have a mass of 4 and a charge of +2
>
> D — deuterons have a mass of 2 and a charge of +1
>
> H — protons have a mass of 1 and a charge of +1

The heavy mass and electrical charge characterize the hazardous properties of alpha particles, deuterons, and protons.

- They are heavy, low velocity particles which penetrate the human body or metal shielding materials only slightly.

- They are positively charged and interact with body materials and metals electrically. They react with electrons (e^-) to form the corresponding neutral atoms.

> proton $+ e^- \rightarrow$ hydrogen atom
>
> deuteron $+ e^- \rightarrow$ deuterium atom
>
> α particle $+ e^- \rightarrow$ helium atom

- They are emitted from radioactive elements at energy levels as high as 8 MeV.

The alpha particles, deuterons, and protons are highly hazardous because of their generally high energy levels. On the other hand, relatively thin shielding suffices to protect the human body from radiation damage. Knowledge of radiation properties permits the easy management of heavy, charged particles by the use of protective shielding.

BETA RAYS

Beta rays, or high energy electrons, are emitted from a wide range of light and heavy radioactive elements. Beta rays are much smaller than alpha particles and may have a velocity approaching the velocity of light. The energy level of beta particles can be as high as 4 MeV. This is the same energy range characteristics of gamma rays.

- The electron is a light particle of high velocity and has great penetrating power.

- The electron has a negative electrical charge which tends to limit its penetration range somewhat.

- Beta rays are relatively more hazardous than alpha rays because the penetration power is greater.

POSITRONS

Positrons are similar to beta rays except that they are positively charged. A positron can be viewed as a positively charged electron.

positron	e^{+}
electron	e^{-}

Positrons are more hazardous than electrons:

- They are emitted from radioactive elements at a similar level as electrons, i.e. as high as 4 MeV.

- They always emit gamma rays when the positron reacts with an electron.

$$e^0_{+1} + e^0_{-1} \rightarrow \gamma^0$$

- The energy level of the gamma ray resulting from the reaction of positrons and electrons is always 0.51 MeV.

Element	Symbol	Isotope	Half-Life, Years	Mode of Decay	
				Radiation	MEV
hydrogen	H	3	12.5	beta	0.02
beryllium	Be	10	2.7×10^6	beta	0.56
carbon	C	14	5700	beta	0.16
sodium	Na	22	2.6	positron	0.5, 1.8
				gamma	1.3
chlorine	Cl	36	4×10^5	beta	0.7
argon	Ar	39	265	beta	0.6
potassium	K	40	1.4×10^9	beta	1.4
				gamma	1.5
cobalt	Co	60	5.25	beta	0.3
				gamma	1.2, 1.3
nickel	Ni	63	85	beta	0.07
selenium	Se	79	6.5×10^4	beta	0.2
krypton	Kr	85	9.4	beta	0.7
				gamma	0.5
rubidium	Rb	87	6×10^{10}	beta	0.3
strontium	Sr	90	25	beta	0.5
zirconium	Zr	93	5×10^6	beta	0.06
palladium	Pd	107	7×10^6	beta	0.04
iodine	I	129	1.7×10^7	beta	0.1
				gamma	0.04
cesium	Cs	135	2.9×10^6	beta	0.2
		137	30	beta	0.5, 1.2
barium	Ba	133	10	gamma	0.09, 0.3
samarium	Sm	147	6.7×10^{11}	alpha	2.1
		151	70	beta	0.08
gadolinium	Gd	150	10^4	alpha	2.7
polonium	Po	208	3	alpha	5.2
radium	Ra	226	1620	alpha	4.8
				gamma	0.2
uranium	U	232	70	alpha	5.3
				gamma	0.06
		233	1.6×10^5	alpha	4.8
				gamma	0.04, 0.08, 0.3
		234	2.5×10^5	alpha	4.8
				gamma	0.06
		235	7.1×10^8	alpha	4.4, 4.6
		236	2.5×10^7	alpha	4.5
				gamma	0.05

Table 36-1. Hazardous Properties of the Radioactive Isotopes Extensively Used in Industry and Medicine.

RADIATION PROPERTIES OF THE MORE FAMILIAR RADIOISOTOPES

The various processes of nuclear fusion reactions, isotopic by-products from nuclear reactors, and isolation of naturally occurring radioisotopes have made available radioactive isotopes of most of the stable elements. The use of radioisotopes in medicine, industry, and in the research laboratory is now commonplace. Although most of these radioactive isotopes are hazardous materials, all can be handled safely if proper precautions are taken. The more commonly used radioisotopes and their associated radiation properties are given in table 36-1.

An important parameter in characterizing radioactive isotopes is the rate at which they decay. The rate at which a radioisotope decays is expressed in terms of the *half-life* or the time during which half of the atoms initially present disintegrate.

- Many radioisotopes have half-lives ranging from fractions of a second to a few hours or days. Most of the short-lived isotopes are not of practical importance as hazardous materials because they decay to stable materials very quickly. Their application is limited to the immediate vicinity of the nuclear reactor where they are made.

- Most of the radioisotopes chosen for use in medicine and industry have half-lives measured in years because of the inherent time delays associated with manufacture, shipping, and storage prior to use. For this reason, isotopes listed in table 36-1 are restricted to those having sufficiently long half-lives to be of practical use.

Radiation Intensity and Dose

Thus far, the discussion of hazardous radioactive materials has emphasized the nature of the radiation resulting from radioactive disintegration processes. Another equally important aspect is the problem of radiation intensity and radiation dose. In relating the measured values of radiation to biological effects produced by it, two distinct types of radiation measurement units have evolved.

- Radiation flux is based on the number of radioactive disintegrations per unit time and is a measure of the quantity of the isotope. Radiation flux is the quantity measured by a Geiger or a scintillation counter.

- Radiation dose measures the quantity of radiation which is absorbed by an object. Radiation dose is of particular value because it is a more direct measure of the biological effect.

Unfortunately, the relationships between radiation flux and dose units are not simple, and one cannot be calculated from a knowledge of the other. All health protection monitoring instruments are designed to measure radiation dose because it is the only quantity which can be related to biological effect. Safety monitoring instruments give scale readings in roentgen units (abbreviated r). A roentgen is a measure of the amount of energy liberated in a unit mass of material. The effects of exposure to various levels of radiation intensity are shown in table 36-2. Survey meters, commonly called geiger counters or scintillation counters (ionization chambers), are used to measure this dose rate.

Dose, Roentgen	Body Effect
0–50	None or minor
100–200	Vomiting and nausea, no deaths
300	Vomiting and nausea followed by radiation sickness, 20% death rate
400–500	50% death rate
750 and above	100% death rate

Table 36-2. Effects of Acute Whole-Body Exposure to Various Radiation Doses.

Three definitions based on the roentgen radiation dose system are commonly encountered. None is entirely satisfactory but each has certain advantages associated with it. It should be emphasized again that the roentgen is the fundamental unit of radiation dosage and those which follow are derived from the roentgen definition.

- REP is an abbreviation for roentgen equivalent physical. One rep is defined as the amount of radiation which will result in the absorption of 93 ergs per gram of tissue. This permits quantitative definition of the different effects of gamma, alpha, beta and other types of radiation.

- REM is an abbreviation for roentgen equivalent man. It is defined as the quantity of radiation which when absorbed in a human body will produce the same biological effect as one roentgen of X-rays.

- RAD is an abbreviation for radiation absorbed dose. It is defined as the equivalent of 100 ergs per gram absorbed in a specific material. When dosage units are expressed in rads the material must be specified also.

RADIATION EXPOSURE

Radiation poisoning or exposure is a rare occurrence for the fire fighter. However, in the event that it does happen, he should be aware of the necessary measures to take to protect himself from continued exposure.

Biological sickness can be measured by the amount of radiation received. It is measured in dosage rates of roentgens per hour (r/hr). For instance, a fire fighter exposed to 10 roentgens in two hours would have a dose rate of 5 r/hr; for 10 roentgens in four hours, he would have a dose rate of 2.5 r/hr.

The presence of radioactive material is always clearly indicated by the familiar purple propeller on a yellow background. This symbol is found wherever radioactive material is present, whether in a research lab or in a transporting vehicle.

Radioactive accidents in a laboratory would most probably be handled by authorized laboratory personnel, and fire fighters would probably not be involved. However, fires could cause rupture or exposure of a radioactive source, thus spewing radioactive contamination throughout the area via smoke and gases. This is the situation that poses danger to fire fighters.

The degree of damage to the body is measured in roentgens per hour. The amount of damage depends on three important factors:

- Length of time exposed.
- Distance from the source of radiation.
- Amount of shielding.

The longer the exposure, the greater chance for radiation injury. This is why it is important to remain at the scene of a radiation accident only for the time absolutely necessary. Distance too, is important. As the distance between the fire fighter and the source increases, the less chance there is for biological damage. Shielding is extremely important. It greatly reduces chances for damage and absorbs dangerous radiation.

The amount of radiation is determined by counter or an ionization chamber. The Geiger counter gives readings in milliroentgens per hour and is a useful means of measuring low level gamma radiation. The ionization chamber is used to measure dosages from 100 milliroentgens.

A fire fighter can tolerate up to 100 r per hour for a short time with no illness. NFPA recommendations, however, establish the safe limit at 25 r. At 200 r, he would become sick; at 300 r, he would become very

REGIONAL COORDINATING OFFICE	POST OFFICE ADDRESS	TELEPHONE FOR ASSISTANCE	D D D AREA CODE
① BROOKHAVEN AREA OFFICE	UPTON, L.I. New York 11973	924-6262	518
② OAK RIDGE OPERATIONS OFFICE	P.O. BOX E OAK RIDGE, TENNESSEE 37830	483-8611, Ext. 3-4510	615
③ SAVANNAH RIVER OPERATIONS OFFICE	P.O. BOX A AIKEN, S.C. 29801	N. AUGUSTA, S.C. 824-6331, Ext. 3333	803
④ ALBUQUERQUE OPERATIONS OFFICE	P.O. BOX 5400 ALBUQUERQUE, NEW MEXICO 87115	264-4667	505
⑤ CHICAGO OPERATIONS OFFICE	9800 S. CASS AVE. ARGONNE, ILLINOIS 60439	739-7711 Ext. 2111 duty hrs. Ext. 4451 off hrs.	312
⑥ IDAHO OPERATIONS OFFICE	P.O. BOX 2108 IDAHO FALLS, IDAHO 83401	526-0111 Ext. 1515	208
⑦ SAN FRANCISCO OPERATIONS OFFICE	2111 BANCROFT WAY BERKELEY, CALIFORNIA 94704	841-5121 Ext. 664 duty hrs. 8419244 off hrs.	415
⑧ RICHLAND OPERATIONS OFFICE	P.O. BOX 550 RICHLAND, WASHINGTON 99352	942-7381	509

U.S. ATOMIC ENERGY COMMISSION
REGIONAL COORDINATING OFFICES
FOR
RADIOLOGICAL ASSISTANCE
AND
GEOGRAPHICAL AREAS
OF RESPONSIBILITY

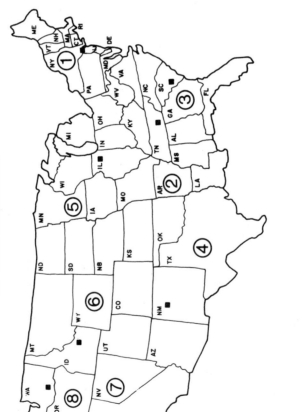

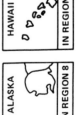 ALASKA — IN REGION 8

 HAWAII — IN REGION 7

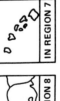 CANAL ZONE — IN REGION 3

 PUERTO RICO — IN REGION 2

 VIRGIN IS. — IN REGION 2

Fig. 36-2.

sick; if exposure was prolonged, he probably then would die. Radiation exposure of 750r or more would cause certain death, depending on exposure limits.

If the fire fighter does get involved with radioactive exposure, he should start decontamination procedures immediately. This involves removing the clothing and showering hair and parts of the body that have been exposed. These should be washed thoroughly. The Atomic Energy Commission (AEC) should be notified in all cases of suspected or actual radiation accidents. The regional locations are shown in figure 36-2.

As mentioned earlier, survey meters (Geiger counters, ionization chambers) measure radioactivity (gamma dose rates) in roentgens per hour (r/hr) or milliroentgens per hour (mr/hr). The smaller counter measures from 0-50 mr/hr and is considered a low range instrument that measures gamma dose rates and detects the presence of beta radiation. Since it is designed for low level measurements, it has limited usefulness in areas of high contamination.

For larger levels of radiation, a 0-500 r/hr counter is available. This unit measures gamma radiation only. These meters are to be used by qualified personnel only, since interpretation of "noise" is important in obtaining readings.

SYMPTOMS OF RADIATION INJURY

Sickness in most cases of radiation exposure is not immediate. The condition and sickness or damage depend on the amount of dose incurred. Some symptoms of radiation sickness are nausea, vomiting, diarrhea, fever, listlessness, and a genuine fatigue feeling. There is no set schedule for the appearance of these symptoms. They may appear in a few days, then even disappear and reappear. If dosage is heavy, the sickness generally appears earlier and is more severe.

Radiation sickness caused by beta radiation received through fallout from particles carried with smoke onto the skin is characterized by itching and burning sensations. Loss of hair may occur. Sores may also appear on the skin. At no time should a fire fighter neglect to notify authorities if he even suspects radiation poisoning.

TRANSPORTATION

The Department of Transportation (DOT) regulates the transportation of radioactive materials and classifies them according to their degree of hazard. The first classification is that of fissile radioactive materials. This includes those radioactive materials that are probably most familiar to the fire fighter — plutonium and uranium.

Three classes are assigned to fissile radioactive materials: Fissile Class I, Fissile Class II and Fissile Class III.

- Fissile Class I designation indicates that the materials being transported require no nuclear criticality safety controls and may be transported in any quantity and in any arrangement.

- Fissile Class II also does not require any nuclear criticality safety controls; provisions are similar to Class I except for limits on the amount of certain materials that may be shipped.

- Fissile Class III shipments require control of nuclear criticality safety during transportation. Special arrangements have to be made between the shipper and the transporter to provide for safe shipment.

For further classification, the DOT has a normal form and a special form for shipping radioactive materials. For the normal form, seven groups are assigned; for the special form only, *special form* is used. For packing the seven normal groups, two types of containers are used — A and B. The difference is that B packages are more accident resistant than A types.

REVIEW

1. Which of the following examples of radiation has the highest energy level?

   a. Ultraviolet light.

   b. Infrared radiation.

   c. Ultrared radiation.

   d. Infraviolet radiation.

2. X-rays are most similar in hazardous properties to

   a. Neutrons.

   b. β-rays.

   c. α-particles.

   d. Ultraviolet radiation.

3. Gamma rays are widely used in medicine because

   a. They are the least hazardous of the fundamental radioactive particles.

   b. They make better photographic plates than neutrons.

   c. They only penetrate a few layers of body tissue.

   d. They can penetrate body tissue and metal.

4. Which of the following forms of radiation is most hazardous?

   a. Ultraviolet.

   b. X-rays.

   c. Infrared.

   d. Light.

5. Gamma rays have high penetration power because they are

   a. Large neutral particles.

   b. Large neutral radiation.

   c. Small neutral radiation.

   d. Small, highly charged rays.

6. All gamma rays

   a. Have the same energy level.

   b. Are equally hazardous.

   c. Are hazardous.

   'd. All of these.

7. Which of the following particles has the highest mass?

   a. Proton.

   b. Deuteron.

   c. Neutron.

   d. Alpha particle.

8. Which of the following has the greatest penetration power?

   a. Proton.

   b. Deuteron.

   c. β-ray.

   d. α-particle.

9. Positrons are more hazardous than electrons because

   a. They are positively charged and therefore have greater penetration power.

   b. They react with electrons to form gamma rays.

   c. They are smaller than electrons.

   d. They react with protons to form antimatter.

10. Most isotopes used in medical applications are chosen with a

    a. Short half-life so that no radioactive by-products remain after use.

    b. Low energy level so that less body damage results.

    c. Long half-life so that the isotope retains radioactive properties after delivery to the user.

    d. Variable half-life so that it can be used for multiple applications.

11. Radiation dose is measured in units of

a. Time.
b. Counts per unit time.
c. Roentgens.
d. Roentgens per unit time.

12. Injuries or fatalities resulting from whole-body exposure to a radiation level of 300 roentgens are

a. Independent of the type of radiation.
b. Greater for gamma rays with high penetration power.
c. Less for α-particles with low penetration.
d. Maximum for beta radiation.

SECTION 10 UNIFYING PRINCIPLES

Unit 37 Official Regulations for Handling Hazardous Materials

Official regulations for hazardous materials are intended to provide guidelines and recommended standard practices for shipping, storage, and handling of a wide variety of materials. The range of materials which must be transported and stored is steadily increasing. The net result is a parallel increase in complexity and specificity of official regulations concerning the handling of these materials.

Many agencies are involved with various aspects of handling hazardous materials, and each publishes official regulations and supplementary documentation of the hazardous properties of materials with which they deal. The various official regulations are all similar but none are the same. They differ primarily in extent of coverage, detail, and intended application.

The **National Fire Protection Association**[1] publishes the *Fire Protection Guide on Hazardous Materials*. As the title indicates, the document is oriented primarily toward fire, explosion and associated accident prevention. It is an important source of basic information and a guide to handling hazardous materials for fire protection personnel. The contents include

- Flash point index of trade name liquids: Lists over 7000 trade name products with flash point data and principal use.

- Fire hazard properties of flammable liquids, gases, and volatile solids: Lists over 1000 flammable materials with data on flash point, ignition temperature, upper and lower flammable limits, specific gravity, vapor density, boiling point, water solubility, and recommended extinguishing method.

- Hazardous chemicals data: Lists supplementary fire, explosion, and toxicity hazards for over 250 chemical materials.

- Manual of Hazardous Chemical Reactions: Lists mixtures of two or more chemicals reported to be potentially hazardous with respect to fire or explosion.

- Recommended system for the identification of the fire hazard of materials: Presents recommended identification system to simplify determination of the degree of health hazard, flammability, reactivity with water, radioactive hazards, and fire control problems.

The **U.S. Coast Guard**[2] publishes the *Chemical Data Guide for Bulk Shipment by Water*. The coverage of hazardous materials is similar but much less extensive than that of the National Fire Protection Association. The intended purpose is to provide guidance for individuals whose duties require decisions in situations involving bulk chemical shipments by water. The contents include

- Physical properties: Lists physical properties including specific gravity, boiling point, vapor pressure, vapor density, freezing point, and water solubility for approximately 130 chemicals.

[1] National Fire Protection Ass'n, International, 470 Atlantic Ave., Boston, Mass. 02210
[2] Department of Transportation, United States Coast Guard, 400 7th St., S.W., Washington, D.C. 20520

- Fire and Explosion Hazard Data: Lists flash point, flammable limits, autoignition temperature, extinguishing media, and special fire procedures for approximately 130 chemicals.

- Health Hazard Data: Lists toxicity rating, toxicity characteristics, symptoms, short exposure tolerance, and exposure procedures for approximately 130 chemicals.

- Reactivity Data: Lists stability and compatibility data for approximately 130 compounds.

- Spill or Leak Data: Lists the immediate steps to be taken should the material be released into air, water, or the enclosed structure of the ship.

The **National Association of Mutual Casualty Companies** publishes the *Handbook of Organic Industrial Solvents*. This guide is similar to that of the National Fire Protection Association but is restricted in coverage to consideration of organic solvents. Among the major data listed are boiling point, freezing point, fire hazard rating, upper and lower explosive limits in air, evaporation rate compared to ether, specific gravity, vapor volume, and threshold toxicity limits.

The primary regulatory agency in the U.S. is the **Hazardous Materials Regulations Board of the Department of Transportation.**[3]   In contrast to the agencies previously discussed which issue only recommendations and guides, the Department of Transportation (DOT) issues official regulations which must be followed. Some recently modified DOT labels are shown in figure 37-1. Old labels were changed to meet a growing demand for conformity. The new labels closely resemble those adopted some time ago by the United Nations for use in international shipping of hazardous materials.

One advantage of using the new labels is that the same ones may be used for all shipping; both domestic and export as well as ground and air.   There are several special labels required by other countries and for special commodities shipped by air.   These are used in addition to the sixteen basic labels illustrated here.

Two additional changes that were proposed for the DOT labeling system which have not yet been implemented are shown in figures 37-2 and 37-3.

- Use of a two-digit hazard identification number facilitates conveying information on multiple hazards of a material.

- Labels are required regardless of quantity of material. The previous 1000 pound minimum is deleted. Placards differing slightly in design from labels are used in the new system to indicate quantities in excess of 1000 pounds.

- Hazard information placards based on the hazard information number are keyed to pages in a manual to be published by the Office of Hazardous Materials. This permits conveying additional potential hazard and immediate action information.

The **Manufacturing Chemists Association**[4] publishes a *Guide to Precautionary Labeling of Hazardous Chemicals*.   This document is a guide prepared for manufacturers of potentially hazardous chemicals to facilitate proper labeling.

The Manufacturing Chemists Association also publishes the *MCA Chem-Card Manual,* which is a quick reference to what to do when a certain material is involved in a fire or is leaking and to action to be taken in case of human exposure to hazardous materials. MCA provides services to the public in the event of

[3] Hazardous Materials Regulations Board, Department of Transportation, 400 7th St., S.W., Washington, D.C. 20590

[4] Manufacturing Chemists Association, 1825 Connecticut Ave., N.W., Washington, D.C. 20009.

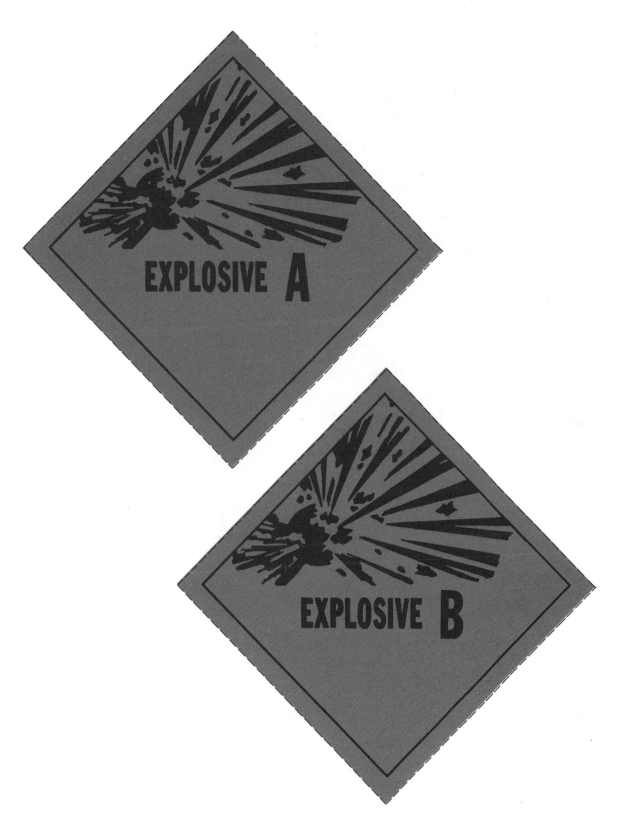

Table 37-1. Modified DOT Labels for Shipping Hazardous Materials (continued).

Table 37-1. Modified DOT Labels for Shipping Hazardous Materials (continued).

Table 37-1. Modified DOT Labels for Shipping Hazardous Materials (continued).

Table 37-1. Modified DOT Labels for Shipping Hazardous Materials (continued).

Table 37-1. Modified DOT Labels for Shipping Hazardous Materials (continued).

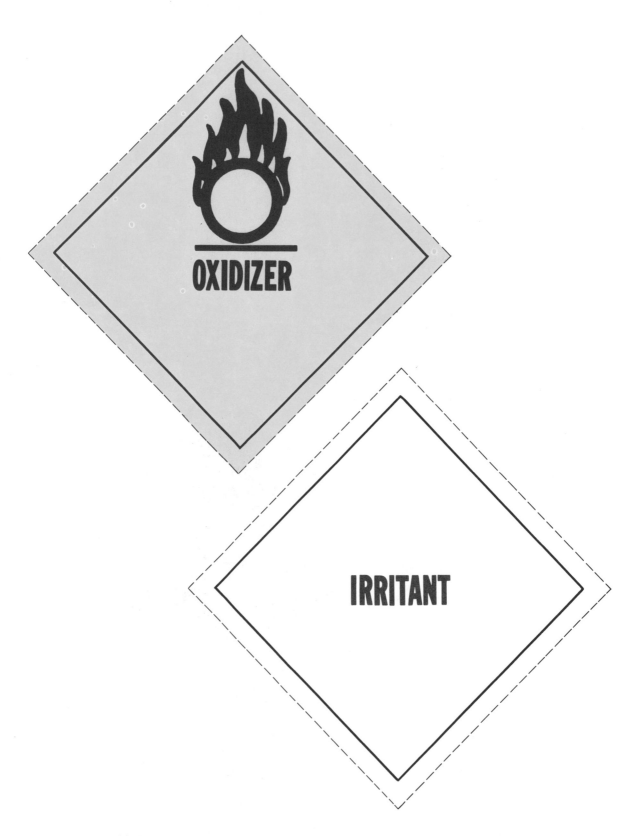

Table 37-1. Modified DOT Labels for Shipping Hazardous Materials (continued).

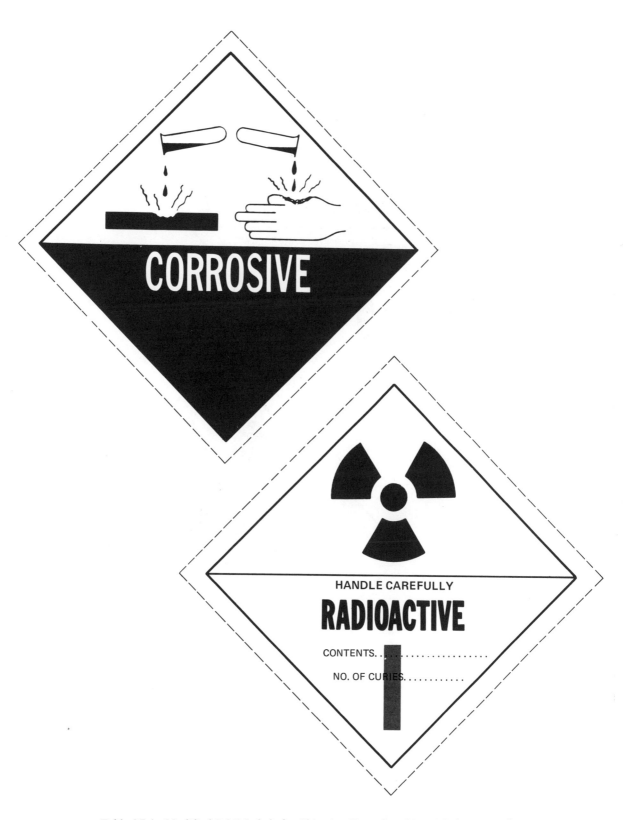

Table 37-1. Modified DOT Labels for Shipping Hazardous Materials (continued).

Table 37-1. Modified DOT Labels for Shipping Hazardous Materials.

Number	Hazard Information	Number	Hazard Information
01	Dangerous	47	Flammable solid, pyrophoric, poison, water reactive
02			
03		48	
04		49	
05	Irritant	50	Oxidizer
06		51	Oxidizer, corrosive
07		52	
08		53	Oxidizer, poison, corrosive
09		54	Oxidizer, thermally unstable
10		55	Oxidizer, thermally unstable, corrosive
11	Explosives Class C	56	Oxidizer, thermally unstable, poison
12		57	Organic peroxide
13		58	Organic peroxide, refrigerated, highly sensitive
14			
15	Explosives Class B	59	Organic peroxide, extremely sensitive
16		60	Poison, highly toxic
17		61	Poison, highly toxic, combustible
18		62	Extremely toxic
19	Explosives Class A	63	
20	Nonflammable gas	64	Extremely or highly toxic by skin absorption
21	Nonflammable gas, corrosive		
22	Oxygen	65	Extremely toxic, flammable or combustible
23	Flammable gas		
24	Flammable gas, corrosive	66	
25		67	Extremely or highly toxic by skin absorption, flammable or combustible
26	Nonflammable gas, poison		
27	Oxidizer gas, poison	68	
28	Flammable gas, poison	69	
29	Flammable gas, poison, extremely hazardous	70	Radioactive, low hazard
30	Combustible or flammable liquid	71	Radioactive
31	Flammable liquid, corrosive	72	Radioactive, oxidizer
32	Flammable liquid, poison	73	Radioactive, corrosive
33		74	Radioactive, pyrophoric
34	Combustible or flammable liquid, self-reactive or thermally unstable	75	
		76	
35	Combustible or flammable liquid, corrosive, self-reactive or thermally unstable	77	
		78	Radioactive, poison, corrosive
36	Combustible or flammable liquid, poison, self-reactive or thermally unstable	79	Radioactive, acid solution of plutonium nitrate
		80	Corrosive
37		81	Corrosive, poison
38	Pyrophoric liquid	82	
39		83	Corrosive, combustible
40	Flammable solid	84	Corrosive, poison, heat of dilution
41	Flammable solid, poison	85	Corrosive, combustible, poison
42	Flammable solid, pyrophoric	86	Corrosive, combustible, heat of dilution
43	Flammable solid, pyrophoric, poison	87	Corrosive, combustible, poison, heat of dilution
44	Flammable solid, water reactive		
45	Flammable solid, poison, water reactive	88	
46	Flammable solid, pyrophoric, water reactive	89	

Table 37-2. Hazard Information Numbers to be Listed in White Block of Proposed New DOT Label.

any transportation emergency involving chemicals through CHEMTREC.[5]

It is apparent that the new DOT shipping regulations have incorporated many of these recommendations. Some typical precautionary statements are summarized in table 37-4. Precautionary measures and instructions in case of contact or exposure are given also, but are not included in the table.

The **Association of American Railroads, Bureau of Explosives,** has a guide entitled

Dangerous Articles Emergency Guide that lists the recommended steps to be taken if there is a fire or a leak. It also gives necessary information about the material involved.

None of the discussions or listing of labels is complete. They are intended to serve as an instruction medium only. Labeling and transportation regulations are always subject to change and the user must be prepared to refer to current, official documents.

---

[5]CHEMTREC (Chemical Transportation Emergency Center) May be reached toll-free day or night, at 800-424-9300, 202-483-7616 (Alaska or Hawaii) and 483-7616 (District of Columbia).

Hazardous Information No. and type	Potential Hazards	Immediate Action Information
01 Dangerous	Contains Hazardous Materials	**General:** No unnecessary personnel. Keep upwind. Identify and isolate hazard area. Wear self-contained breathing apparatus and full-protective clothing. **Fire:** On small fires, use dry chemical or carbon dioxide. On large fires, use standard fire fighting agents. Cool containers with water from maximum distance. Continue cooling after fires have been extinguished. Move exposed containers from fire area, if without risk. **Spill or Leak:** Within hazard area: Eliminate ignition sources. No flares, no smoking, no open flames. Stop leak, if without risk. Use water spray to reduce vapors. Use noncombustible absorbent material (sand, etc.) to collect small spills. Dike large spills for later disposal. **First Aid:** Remove to fresh air. Call physician. If not breathing, give artificial respiration. If breathing is difficult, give oxygen. In case of contact with material or water solution, immediately flush skin or eyes with running water for at least 15 minutes. Remove contaminated clothing and shoes. Keep patient at rest.
05 Irritant	**Health:** Gas very irritating, if breathed. May cause extreme burning of the eyes resulting in a copious flow of tears. May also cause coughing, difficult breathing, and nausea. If exposure is brief, effects last only a few minutes. Effects may be serious if exposed to product in an enclosed unventilated area.	**First Aid:** Remove to fresh air. Remove contaminated clothing. In case of contact with material, immediately flush skin or eyes with running water for at least 15 minutes. Effects should disappear after individual has been exposed to fresh air for 5-10 minutes.

Table 37-3. Summary of Potential Hazard and Immediate Action Information to be Shown in the HI Manual (cont'd).

Hazardous Information No. and Type	Poatential Hazards	Immediate Action Information
11 Explosives C	**Fire:** May burn very rapidly. **Explosion:** Individual items may explode when subjected to heat or fire. **Health:** Fire may produce irritating gases.	**General:** No unnecessary personnel. Identify and isolate hazard area. Self-contained breathing apparatus should be available. Except under emergency conditions, explosives should be handled only under the supervision of an expert. **Fire:** On fire not in cargo area, use unmanned hose holder or monitor nozzles from maximum distance or behind barrier. Withdraw from hazard area if fire advanced or massive in cargo area. **Spill or Leak:** Within hazard area: Eliminate ignition sources. No flares, no open flames, no smoking. **First Aid:** Use standard first aid procedures.
15 Explosives B	**Fire:** May burn very rapidly. Fire very difficult to extinguish by conventional methods.	**General:** No unnecessary personnel. Keep upwind. Identify and isolate hazard area. Self-contained breathing apparatus should be available. Wear full-protective clothing. Except under emergency conditions, explosives should be handled only under the supervision of an expert. **Fire:** If fire not in cargo area, extinguish by conventional methods. If fire in cargo area, use unmanned hose holder or monitor nozzles from maximum distance or behind barrier. Continue cooling after flames have been extinguished. If cargo must be moved, use extreme care in handling and evacuate unnecessary personnel from area. Withdraw from hazard area, if fire advanced or massive in cargo area. **Spill or Leak:** Within hazard area: Eliminate ignition sources. No flares, no open flames, no smoking. Avoid contact with spilled material. **First Aid:** Use standard first aid procedures.
19 Explosives A	**Explosion:** May detonate violently if subjected to heat, flame, or shock. Probability of explosion increases when heated. **Health:** Fire may produce irritating gases.	**General:** No unnecessary personnel. Identify and isolate hazard area. (Recommended minimum radius 1/2 mile.) Except under emergency conditions, explosives should be handled only under the supervision of an expert. **Fire:** If fire or heat from fire not in cargo area, use unmanned hose holder or monitor nozzle from maximum distance or behind barrier, or if without risk, use conventional methods. Do not move cargo or vehicle if cargo has been exposed to fire or heat. If cargo must be moved, use extreme care in handling to avoid shocks, and evacuate unnecessary personnel from area. Continue cooling after flames have been extinguished. Do not fight fire in cargo area, and withdraw from hazard area.

Table 37-3. Summary of Potential Hazard and Immediate Action Information to be Shown in the HI Manual(cont'd).

Hazardous Information No. and Type	Potential Hazards	Immediate Action Information
19 (cont'd)		**Spill or Leak:** Within immediate area: Eliminate ignition sources. No flares, no open flames, no smoking. Avoid contact with spilled material. **First Aid:** Use standard first aid procedures.
20 Nonflammable gas	**Fire:** Some material in this group cannot catch fire, others can catch fire but do not ignite readily. Heated container may rupture violently and produce flying missiles. **Health:** Little or no hazard except in confined areas. Vapors may cause dizziness or suffocation, if breathed.	**General:** No unnecessary personnel. Identify and isolate hazard area. Self-contained breathing apparatus should be available. **Fire:** Move exposed containers from fire area, if without risk. Cool containers with water. Do not approach ends of horizontal tanks. Use standard fire fighting agents. **Spill or Leak:** Avoid contact with liquefied gas. Stop leak if without risk. **First Aid:** Remove to fresh air. Use standard first aid procedures.
21 Nonflammable gas, corrosive	**Fire:** Some material in this group cannot catch fire, others can catch fire but do not ignite readily. Heated container may rupture violently and produce flying missiles. **Health:** Gas very irritating, if breathed. Contact with material may cause severe burns to skin and eyes.	**General:** No unnecessary personnel. Keep upwind. Identify and isolate hazard area. Wear self-contained breathing apparatus and full-protective clothing. **Fire:** Move exposed containers from fire area, if without risk. Cool containers with water. Do not approach ends of horizontal tanks. Use standard fire fighting agents. **Spill or Leak:** Stop leak if without risk. Use water spray to reduce vapors. Keep area isolated until gas has dispersed. **First Aid:** Remove to fresh air. Call physician. If not breathing, give artificial respiration. In case of contact with material or water solution, immediately flush skin or eyes with running water for at least 15 minutes. Remove contaminated clothing and shoes. Keep patient at rest.
22 Oxygen	**Fire:** Materials that do not burn in air may be ignited in oxygen vapor. Reaction with fuel may be violent. **Explosion:** Mixture with fuels may explode. Vapor entering sewers or other closed spaces may create explosion hazard. **Health:** Contact with liquid or cold gas may cause severe skin and eye injury similar to a burn.	**General:** No unnecessary personnel. Identify and isolate hazard area. Wear full-protective clothing. **Fire:** Move exposed containers from fire area, if without risk. Use standard fire fighting agent. Cool containers with water from maximum distance. Use water from maximum distance to protect surrounding property. Withdraw from hazard area, if fire in cargo area is massive or advanced.

Table 37-3. Summary of Potential Hazard and Immediate Action Information to be Shown in the HI Manual (cont'd).

Hazardous Information No. and Type	Potential Hazards	Immediate Action Information
22 (cont'd)		**Spill or Leak:** Within hazard area: Eliminate ignition source. No flares, no smoking, no open flames. Stop leak if without risk. Avoid contact with spilled material. Keep spilled material away from combustibles. Keep area isolated until gas has dispersed. **First Aid:** Remove to fresh air. Call physician. Remove contaminated clothing and shoes. Thaw frosted parts with water. Use standard first aid procedures.
23 Flammable gas	**Fire:** May be ignited by heat, sparks, or open flames. Ignition of vapor may occur at some distance from leaking container. Heated container may rupture violently and produce flying missiles. Vapor entering sewers or other closed spaces may create fire or explosion hazard. **Explosion:** May form explosive mixtures with air. **Health:** Vapors may cause dizziness or suffocation, if breathed.	**General:** No unnecessary personnel. Keep upwind. Identify and isolate hazard area. Self-contained breathing apparatus should be available. **Fire:** Do not approach ends of horizontal tanks. Do not extinguish fire unless leak can be stopped. Use standard fire fighting agents. Cool containers with water. Move exposed containers from fire area, if without risk. If fire in cargo area is massive or advanced, use unmanned hose holder or monitor nozzles. If this is impossible, withdraw from area and let fire burn. **Spill or Leak:** Within hazard area: Eliminate ignition source. No flares, no smoking, no open flames. Stop leak if without risk. Use water spray to reduce vapors. Keep area isolated until gas has dispersed. **First Aid:** Remove to fresh air. Use standard first aid procedures.
24 Flammable gas, corrosive	**Fire:** May be ignited by heat, sparks, or open flames. Ignition of vapor may occur at some distance from leaking container. Heated container may rupture violently and produce flying missiles. Vapor entering sewers or other closed spaces may create fire or explosion hazard. **Explosion:** May form explosive mixtures with air. **Health:** Gas very irritating, if breathed. Contact with material may cause severe burns to skin and eyes.	**General:** No unnecessary personnel. Keep upwind. Identify and isolate hazard area. Wear self-contained breathing apparatus and full-protective clothing. **Fire:** Do not extinguish fire unless leak can be stopped. Use standard fire fighting agents. Do not approach ends of horizontal tanks. Cool containers with water. Move exposed containers from fire area if without risk. If fire in cargo area is massive or advanced, use unmanned hose holder or monitor nozzles. If this is impossible, withdraw from area and let fire burn. **Spill or Leak:** Within hazard area: Eliminate ignition source. No flares, no smoking, no open flames. Stop leak if without risk. Use water spray to reduce vapors. Keep area isolated until gas has dispersed.

Table 37-3. Summary of Potential Hazard and Immediate Action Information to be Shown in the HI Manual (cont'd).

Hazardous Information No. and Type	Potential Hazards	Immediate Action Information
24 (cont'd)		**First Aid:** Remove to fresh air. Call physician. If not breathing, give artificial respiration. If breathing is difficult, give oxygen. Remove contaminated clothing and shoes. In case of contact with material or water solution, immediately flush skin or eyes with running water for at least 15 minutes. Use standard first aid procedures. Keep patient at rest.
26 Nonflammable gas, poison	**Fire:** Heated container may rupture violently and produce flying missiles. Some material in this group cannot catch fire, others catch fire but do not ignite readily. **Health:** Vapor is poisonous, can be fatal if breathed in high concentrations. Contact with material may cause severe burns to skin and eyes. Contaminated water or material runoff may pollute water supply.	**General:** No unnecessary personnel. Keep upwind. Identify and isolate hazard area. Wear self-contained breathing apparatus and full-protective clothing. **Fire:** Evacuate where poison gas may endanger inhabited area. Do not approach ends of horizontal tanks. Cool containers with water. Move exposed containers from fire area, if without risk. Use standard fire fighting agents. **Spill or Leak:** Evacuate where poison gas may endanger inhabited area. Stop leak if without risk. Use water spray to reduce vapors. Keep area isolated until gas has dispersed. **First Aid:** Remove to fresh air. Call physician. If not breathing, give artificial respiration. If breathing is difficult, give oxygen. Remove contaminated clothing and shoes. In case of contact with material or water solution, immediately flush skin or eyes with running water for at least 15 minutes. Keep patient at rest. Effects of contact or inhalation may be delayed.
27 Oxidizer, poison	**Fire:** May cause fire on contact with combustibles. Reaction with fuels may be violent. Vapors entering sewers or other closed spaces may create fire or explosion hazard. Heated container may rupture violently and produce flying missiles. **Explosion:** Mixture with fuels may explode. **Health:** Vapor is poisonous, can be fatal if breathed in high concentrations. Contact with material may cause severe burns to skin and eyes. Contaminated water or material runoff may pollute water supply.	**General:** No unnecessary personnel. Keep upwind. Identify and isolate hazard area. Wear self-contained breathing apparatus and full-protective clothing. **Fire:** Evacuate where poison gas may endanger inhabited area. Cool containers with water. On small fires, use dry chemical. On large fires, use water, spray, or fog. Use water spray to protect surrounding area.

Table 37-3. Summary of Potential Hazard and Immediate Action Information to be Shown in the HI Manual (cont'd).

Hazardous Information No. and Type	Potential Hazards	Immediate Action Information
27 (cont'd)		**Spill or Leak:** Evacuate where poison gas may endanger inhabited area. Stop leak if without risk. Avoid contact with spilled material. Keep spilled material away from combustibles. Use water spray to reduce vapors. Dilute spill with large amounts of water. Dike for later disposal. Keep area isolated until gas has dispersed. **First Aid:** Remove to fresh air. Call physician. If not breathing, give artificial respiration. If breathing is difficult, give oxygen. In case of contact with material or water solution, immediately flush skin or eyes with running water for at least 15 minutes. Remove contaminated clothing and shoes. Keep patient at rest.
28 Flammable gas, poison	**Fire:** May be ignited by heat, sparks, or open flames. Ignition of vapor may occur at some distance from leaking container. Heated container may rupture violently and produce flying missiles. Vapor entering sewers or other closed spaces may create fire, explosion, or poison hazard. **Explosion:** May form explosive mixture with air. **Health:** Vapor is poisonous, can be fatal if breathed in high concentrations. Contact with material may cause severe burns to skin and eyes. Contaminated water or material runoff may pollute water supply.	**General:** No unnecessary personnel. Keep upwind. Identify and isolate hazard area. Wear self-contained breathing apparatus and full-protective clothing. **Fire:** Evacuate where poison gas may endanger inhabited area. Do not extinguish fire unless leak can be stopped. Do not approach ends of horizontal tanks. Cool containers with water. Move exposed containers from fire area, if without risk. Use standard fire fighting agents. **Spill or Leak:** Within hazard area: Eliminate ignition source. No flares, no smoking, no open flames. Evacuate where poison gas may endanger inhabited area. Stop leak if without risk. Use water spray to reduce vapors. Keep area isolated until gas has dispersed. **First Aid:** Remove to fresh air. Call physician. If not breathing, give artificial respiration. If breathing is difficult, give oxygen. In case of contact with material or water solution, immediately flush skin or eyes with running water for at least 15 minutes. Remove contaminated clothing and shoes. Keep patient at rest. Effects of contact or inhalation may be delayed.
29 Flammable gas, poison, extremely hazardous	**Fire:** Some gases in this group are easily ignited by heat, sparks, or open flames; others can catch fire but do not ignite readily, or may react violently with combustibles. Heated container may rupture violently and produce flying missiles.	**General:** No unnecessary personnel. Keep upwind. Identify and isolate hazard area. Wear self-contained breathing apparatus and full-protective clothing.

Table 37-3. Summary of Potential Hazard and Immediate Action Information to be Shown in the HI Manual (cont'd).

Hazardous Information No. and Type	Potential Hazards	Immediate Action Information
29 (cont'd)	Vapor entering sewers or closed spaces may create poison, explosion, or fire hazard. **Health:** Vapor very poisonous. Breathing of vapor causes little irritation. Fatal concentrations can be readily inhaled without noticing. Runoff may pollute water supply.	**Fire:** Use water spray or fog. Fight fire from maximum distance and from behind barrier. Use water from maximum distance and from behind barrier. Evacuate where poison gas may endanger inhabited area. Do not extinguish fire unless gas flow can be stopped. **Spill or Leak:** Within hazard area: Eliminate ignition source. No flares, no smoking, no open flames. Evacuate where poison gas may endanger inhabited area. Stop leak if without risk. Use water spray to reduce vapors. Keep area isolated until gas has dispersed. **First Aid:** Remove to fresh air. Call physician. If not breathing, give artificial respiration. If breathing is difficult, give oxygen. Remove contaminated clothing and shoes. In case of contact with material or water solution, immediately flush skin or eyes with running water for at least 15 minutes. Keep patient at rest. Effects of contact or inhalation may be delayed.
30 Combustible or Flammable liquid	**Fire:** May be ignited by heat or open flames. Heated container may rupture violently and produce flying missiles. Ignition of vapor may occur at some distance from leaking container. Vapor entering sewer or other closed spaces may create explosion hazard. **Health:** Fire may produce irritating gases. Vapors may cause dizziness or suffocation, if breathed.	**General:** No unnecessary personnel. Keep upwind. Identify and isolate hazard area. Self-contained breathing apparatus should be available. **Fire:** On small fires, use dry chemical or carbon dioxide. On large fires, use standard fire fighting agents. Do not approach ends of horizontal tanks. Move exposed containers from fire area, if without risk. Cool containers with water. Continue cooling after fire has been extinguished. **Spill or Leak:** Within hazard area: Eliminate ignition source. No flares, no smoking, no open flames. Stop leak if without risk. Use water spray to reduce vapors. Dike large spills for later disposal. Use noncombustible absorbent material such as sand to collect small spills. **First Aid:** Remove to fresh air. Use standard first aid procedures.
31 Flammable liquid, corrosive	**Fire:** May be ignited by heat, sparks, or open flames. Heated container may rupture violently and produce flying missiles. Ignition of vapor may occur at some distance from leaking container.	**General:** No unnecessary personnel. Keep upwind. Identify and isolate hazard area. Self-contained breathing apparatus should be available.

Table 37-3. Summary of Potential Hazard and Immediate Action Information to be Shown in the HI Manual (cont'd).

Hazardous Information No. and Type	Potential Hazards	Immediate Action Information
31 (cont'd)	Vapor entering sewers or other closed spaces may create fire explosion hazards. **Health:** Fire may produce irritating gases. Vapor may be irritating, if breathed. Contact with material may cause severe burns to skin and eyes. Contaminated water or material runoff may pollute water supply.	**Fire:** On small fires, use dry chemical or carbon dioxide. On large fires, standard fire fighting agents are used. Do not approach ends of horizontal tanks. Cool containers with water. Continue cooling after fires have been extinguished. Move exposed containers from fire area, if without risk. **Spill or Leak:** Within hazard area: Eliminate ignition sources. No flares, no smoking, no open flames. Stop leak if without risk. Use water spray to reduce vapors. Dike large spills for later disposal. Use noncombustible absorbent material such as sand to collect small spills. **First Aid:** Remove to fresh air. Call physician. In case of contact with material or water solution, immediately flush skin or eyes with running water for at least 15 minutes. Remove contaminated clothing and shoes. Keep patient at rest.
32 Flammable liquid, poison	**Fire:** May be ignited by heat, sparks, or open flames. Heated container may rupture violently and produce flying missiles. Ignition of vapor may occur at some distance from leaking container. Vapor entering sewers or other closed spaces may create fire or explosion hazard. **Health:** Vapor is poisonous if breathed. Contact with material may cause severe burns to skin and eyes. Liquid may cause death, if consumed. Runoff may pollute water supply.	**General:** No unnecessary personnel. Keep upwind. Identify and isolate hazard area. Wear self-contained breathing apparatus and full-protective clothing. **Fire:** On small fires, use dry chemical or carbon dioxide. On large fires, standard fire fighting agents are used. Cool containers with water from maximum distance. Continue cooling after fires have been extinguished. Do not approach ends of horizontal tanks. Move exposed containers from fire area, if without risk. **Spill or Leak:** Within hazard area: Eliminate ignition sources. No flares, no smoking, no open flames. Stop leak, if without risk. Use water spray to reduce vapors. Use noncombustible absorbent material (sand) to collect small spills. Dike large spills for later disposal. **First Aid:** Remove to fresh air. Call physician. If not breathing, give artificial respiration. If breathing is difficult, give oxygen. In case of contact with material or water solution, immediately flush skin or eyes with running water for at least 15 minutes. Remove contaminated clothing and shoes. Keep patient at rest. Effects of contact or inhalation may be delayed.
34 Flammable liquid, self-reactive	**Fire:** May be ignited by heat, sparks, or open flames. Ignition	**General:** No unnecessary personnel. Keep upwind. Identify and isolate hazard area. Wear full-protective

Table 37-3. Summary of Potential Hazard and Immediate Action Information to be Shown in the HI Manual (cont'd).

Hazardous Information No. and Type	Potential Hazards	Immediate Action Information
34 or thermally unstable	of vapor may occur at some distance from leaking container. Heated container may rupture violently and produce flying missiles even if water is applied for cooling. Vapor entering sewers or other closed spaces may create fire or explosion hazard. **Health:** Contact with material may cause severe burns to skin and eyes. Fire may produce poison gases. Contaminated water or material runoff may pollute water supply.	clothing. Self-contained breathing apparatus should be available. **Fire:** Cool containers with water from maximum distance. Continue cooling after fires have been extinguished. Do not approach ends of horizontal tanks. On small fires, use dry chemical or carbon dioxide. On large fires, use standard fire fighting agents. Move exposed containers from fire area, if without risk. Withdraw from hazard area in case of rising sound from venting safety device. If fire in cargo area is massive or advanced, withdraw from hazard area and use unmanned hose holder or monitor nozzle. **Spill or Leak:** Within hazard area: Eliminate ignition source. No flares, no smoking, no open flames. Stop leak if without risk. Use water spray to reduce vapors. Dike large spills for later disposal. **First Aid:** Remove to fresh air. Use standard first aid procedures.
35 Flammable liquid, corrosive, self-reactive or thermally unstable	**Fire:** May be ignited by heat, sparks, or open flame. Ignition of vapor may occur at some distance from leaking container. Heated container may rupture violently and produce flying missiles even if water is applied for cooling. **Health:** Contact with material may cause severe burns to skin and eyes. Fire may produce poison gases. Contaminated water or material runoff may pollute water supply.	**General:** No unnecessary personnel. Keep upwind. Identify and isolate hazard area. Wear full-protective clothing. Self-contained breathing apparatus should be available. **Fire:** Cool containers with water from maximum distance. Continue cooling after fires have been extinguished. Do not approach ends of horizontal tanks. On small fires, use dry chemical or carbon dioxide. On large fires, use standard fire fighting agents. Move exposed containers from fire area, if without risk. Withdraw from hazard area in case of rising sound from venting safety device. If fire in cargo area is massive or advanced, withdraw from hazard area and use unmanned hose holder or monitor nozzles. **Spill or Leak:** Within hazard area: Eliminate ignition source. No flares, no smoking, no open flames. Stop leak if without risk. Use water spray to reduce vapors. Dike large spills for later disposal. **First Aid:** Remove to fresh air. Call physician. In case of contact with material or water solution, immediately flush skin or eyes with running water for at least 15 minutes. Remove contaminated clothing and shoes. Keep patient at rest. Effects of contact or inhalation may be delayed.

Table 37-3. Summary of Potential Hazard and Immediate Action Information to be Shown in the HI Manual (cont'd).

Hazardous Information No. and Type	Potential Hazards	Immediate Action Information
36 Flammable liquid, poison, self-reactive or thermally unstable	**Fire:** May be ignited by heat, sparks, or open flame. Ignition of vapor may occur at some distance from leaking container. Heated container may rupture violently and produce flying missiles even if water is applied for cooling. Vapor entering sewers or other closed spaces may create fire or explosion hazard. **Health:** Vapor is poisonous, if breathed. Liquid or solid may cause death, if consumed. Fire may produce poisonous gases. Contact with material may cause severe burns to skin and eyes. Contaminated water or material runoff may pollute water supply.	**General:** No unnecessary personnel. Keep upwind. Identify and isolate hazard area. Wear self-contained breathing apparatus and full-protective clothing. **Fire:** Cool containers with water from maximum distance. Continue cooling after fires have been extinguished. Do not approach ends of horizontal tanks. On small fires, use dry chemical or carbon dioxide. On large fires, use standard fire fighting agents. Move exposed containers from fire area, if without risk. Withdraw from hazard area in case of rising sound from venting safety device. If fire in cargo area is massive or advanced, withdraw from hazard area and use unmanned hose holder or monitor nozzles. **Spill or Leak:** Within hazard area: Eliminate ignition source. No flares, no smoking, no open flames. Use water spray to reduce vapors. Dike large spills for later disposal. **First Aid:** Remove to fresh air. Call physician. If not breathing, give artificial respiration. If breathing is difficult, give oxygen. In case of contact with material or water solution, immediately flush skin or eyes with running water for at least 15 minutes. Remove contaminated clothing and shoes. Keep patient at rest. Effects of contact or inhalation may be delayed.
38 Pyrophoric liquid	**Fire:** Burns very rapidly and intensely, sometimes with flare-burning effect. May reignite after fire is extinguished. May catch fire spontaneously in air. May react with water to release flammable gas. Runoff to sewer may create fire or explosion hazard. **Health:** Contact with material may cause severe burns to skin or eyes.	**General:** No unnecessary personnel. Keep upwind. Identify and isolate hazard area. Self-contained breathing apparatus should be available. **Fire:** On small fires, use dry chemical or carbon dioxide. On large fires, use flooding amounts of water. Move exposed containers from fire area, if without risk. Continue cooling after fires have been extinguished. Use water from maximum distance to protect surrounding property. If fire in cargo area is massive or advanced, withdraw from hazard area and use unmanned hose holder or monitor nozzles. **Spill or Leak:** Stop leak if without risk. Cover small spills with dry sand or similar inert absorbent. Collect into clean, dry metal container and keep tightly covered. Dilute spill with large amounts of water. Dike for later disposal. **First Aid:** Remove to fresh air. Call physician. Remove contaminated clothing and shoes. In case of contact with material or water solution, immediately

Table 37-3. Summary of Potential Hazard and Immediate Action Information to be Shown in the HI Manual (cont'd).

Hazardous Information No. and Type	Potential Hazards	Immediate Action Information
38 (cont'd)		flush skin or eyes with running water for at least 15 minutes. Keep patient at rest.
40 Flammable solid	**Fire:** Burns very rapidly and intensely, sometimes with flare-burning effect. May be ignited by heat, sparks, or open flame. **Health:** Contact with material may cause severe burns to skin and eyes.	**General:** No unnecessary personnel. Keep upwind. Identify and isolate hazard area. Wear full-protective clothing. **Fire:** On small fires, use dry chemical. On large fires, use standard fire fighting agents. Move exposed containers from fire area, if without risk. Cool containers with water from maximum distance. If fire in cargo area is massive or advanced, withdraw from hazard area and use unmanned hose holder or monitor nozzles. **Spill or Leak:** Within hazard area: Eliminate ignition source. No flares, no smoking, no open flames. Stop leak if without risk. Collect into clean, dry metal container and keep tightly covered. Flush small spill area with water spray. **First Aid:** Call physician. Use standard first aid procedures. In case of contact with material or water solution, immediately flush skin or eyes with running water for at least 15 minutes.
41 Flammable solid, poison	**Fire:** Burns very rapidly and intensely, sometimes with flare-burning effect. **Health:** Vapor, mist, or dust is poisonous, if breathed. Contact with material may cause severe burns. Fire may produce poison gases.	**General:** No unnecessary personnel. Keep upwind. Identify and isolate hazard area. Wear self-contained breathing apparatus. **Fire:** On small fires, use dry chemical. On large fires, use standard fire fighting agents. Move exposed containers from fire area, if without risk. Cool containers with water from maximum distance. If fire in cargo area is massive or advanced, withdraw from hazard area and use unmanned hose holder or monitor nozzles. **Spill or Leak:** Within hazard area: Eliminate ignition source. No flares, no smoking, no open flames. Stop leak if without risk. Collect small spills with dry sand or similar inert absorbent. Flush small spill area with water spray. **First Aid:** Remove to fresh air. Call physician. If not breathing, give artificial respiration. If breathing is difficult, give oxygen. In case of contact with material or water solution, immediately flush skin or eyes with running water for at least 15 minutes. Remove contaminated clothing and shoes. Keep patient at rest.

Table 37-3. Summary of Potential Hazard and Immediate Action Information to be Shown in the HI Manual (cont'd).

Hazardous Information No. and Type	Potential Hazards	Immediate Action Information
42 Flammable solid, pyrophoric	**Fire:** May catch fire spontaneously in air. May reignite after fire is extinguished. **Health:** Materials have little health hazard. Contact with material may cause severe burns to skin and eyes.	**General:** No unnecessary personnel. Keep upwind. Identify and isolate hazard area. Self-contained breathing apparatus should be available. **Fire:** On small fires, use dry chemical, sand, or earth. On large fires, use flooding amounts of water. Move exposed containers from fire area, if without risk. Cool containers with water from maximum distance. Continue cooling after fires have been extinguished. If fire in cargo area is massive or advanced, withdraw from hazard area and use unmanned hose holder or monitor nozzles. **Spill or Leak:** Within hazard area: Eliminate ignition source. No flares, no smoking, no open flames. Stop leak if without risk. Cover small spills with dry sand or similar inert absorbent. Collect into clean, dry metal container and keep tightly covered. Flush small spill area with water spray. **First Aid:** Call physician. In case of contact with material or water solution, immediately flush skin or eyes with running water for at least 15 minutes. Remove contaminated clothing and shoes.
43 Flammable solid, pyrophoric, poison	**Fire:** May catch fire spontaneously in air. May reignite after fire is extinguished. **Health:** Vapor, mist or dust is poisonous, if breathed. Contact with material may cause severe burns. Fire may produce poison gases.	**General:** No unnecessary personnel. Keep upwind. Identify and isolate hazard area. Wear self-contained breathing apparatus. **Fire:** On small fires, use dry chemical, sand, or earth. On large fires, use flooding amounts of water. Move exposed containers from fire area, if without risk. Cool containers with water from maximum distance. Continue cooling after fires have been extinguished. If fire in cargo area is massive or advanced, withdraw from hazard area and use unmanned hose holder or monitor nozzles. **Spill or Leak:** Within hazard area: Eliminate ignition source. No flares, no smoking, no open flames. Stop leak if without risk. Cover small spills with dry sand or similar inert absorbent. Collect into clean, dry metal container and keep tightly covered. Flush small spill area with water spray. **First Aid:** Remove to fresh air. Call physician. If not breathing, give artificial respiration. If breathing is difficult, give oxygen. In case of contact with material or water solution, immediately flush skin or eyes with running water for at least 15 minutes. Speed in

Table 37-3. Summary of Potential Hazard and Immediate Action Information to be Shown in the HI Manual (cont'd).

Hazardous Information No. and Type	Potential Hazards	Immediate Action Information
43 (cont'd)		removing material from skin is of extreme importance. Keep patient at rest. Effects of contact or inhalation may be delayed.
44 Flammable solid, water reactive	**Fire:** Burns very rapidly and intensely, sometimes with flare-burning effect. May catch fire spontaneously in air. May react with water to release flammable gas. **Health:** Contact with material may cause severe burns.	**General:** No unnecessary personnel. Keep upwind. Identify and isolate hazard area. Self-contained breathing apparatus should be available. **Fire:** On small fires, use dry chemical, sand, or earth. DO NOT USE WATER. Move exposed containers from fire area, if without risk. Let large fire burn. Use water from maximum distance to protect surrounding property. Do not get water inside containers. Withdraw from hazard area, if fire in cargo area is massive or advanced. **Spill or Leak:** With hazard area: Eliminate ignition source. No flares, no smoking, no open flames. Stop leak if without risk. Collect into clean, dry metal container and keep tightly covered. **First Aid:** Call physician. In case of contact with material or water solution, immediately flush skin or eyes with running water for at least 15 minutes. Remove contaminated clothing and shoes.
45 Flammable solid, poison, water reactive	**Fire:** Burns very rapidly and intensely, sometimes with flare-burning effect. May catch fire spontaneously in air. May react with water to release flammable gas. **Health:** Vapor, mist, or dust is poisonous, can be fatal if breathed in high concentrations. Contact with material may cause severe burns. Fire may produce poison gases.	**General:** No unnecessary personnel. Keep upwind. Identify and isolate hazard area. Wear self-contained breathing apparatus. **Fire:** On small fires, use dry chemical, sand, or earth. Move exposed containers from fire area, if without risk. DO NOT USE WATER. Let large fire burn. Use water from maximum distance to protect surrounding property. Withdraw from hazard area, if fire in cargo area is massive or advanced. **Spill or Leak:** Stop leak if without risk. Collect into clean, dry metal container and keep tightly covered. **First Aid:** Remove to fresh air. Call physician. If not breathing, give artificial respiration. If breathing is difficult, give oxygen. In case of contact with material or water solution, immediately flush skin or eyes with running water for at least 15 minutes. Remove contaminated clothing and shoes. Keep patient at rest.
46 Flammable solid,	**Fire:** May catch fire spontaneously in air. May react with water to release flammable gas. May reignite	**General:** No unnecessary personnel. Keep upwind. Identify and isolate hazard area. Self-contained breathing apparatus should be available.

Table 37-3. Summary of Potential Hazard and Immediate Action Information to be Shown in the HI Manual (cont'd).

Hazardous Information No. and Type	Potential Hazards	Immediate Action Information
46 pyrophoric, water reactive	after fire is extinguished. Burns very rapidly and intensely, sometimes with flare-burning effect. **Health:** Contact with material may cause severe burns.	**Fire:** On small fires, use dry chemical, sand, or earth. Move exposed containers from fire area, if without risk. DO NOT USE WATER. Let large fire burn. Use water from maximum distance to protect surrounding property. Cool containers with water from maximum distance. Continue cooling after fires have been extinguished. Withdraw from hazard area, if fire in cargo area is massive or advanced. **Spill or Leak:** Stop leak if without risk. Cover small spills with dry sand or similar inert absorbent. Collect into clean, dry metal container and keep tightly covered. **First Aid:** Call physician. In case of contact with material or water solution, immediately flush eyes or skin with running water for at least 15 minutes. Remove contaminated clothing and shoes.
47 Flammable solid, pyrophoric, poison, water reactive	**Fire:** May catch fire spontaneously in air. May react with water to release flammable gas. May reignite after fire is extinguished. Burns very rapidly and intensely, sometimes with flare-burning effect. **Health:** Vapor, mist, or dust is poisonous, can be fatal if breathed in high concentrations. Contact with material may cause severe burns. Fire may produce poison gases.	**General:** No unnecessary personnel. Keep upwind. Identify and isolate hazard area. Wear self-contained breathing apparatus. **Fire:** On small fires, use dry chemical, sand, or earth. Move exposed containers from fire area, if without risk. DO NOT USE WATER. Let large fire burn. Use water from maximum distance to protect surrounding property. Cool container with water from maximum distance. Continue cooling after fires have been extinguished. **Spill or Leak:** Wear self-contained breathing apparatus. Stop leak if without risk. Cover small spills with dry sand or similar inert absorbent. Collect into clean, dry metal container and keep tightly covered. **First Aid:** Remove to fresh air. Call physician. If not breathing, give artificial respiration. If breathing is difficult, give oxygen. In case of contact with material or water solution, immediately flush skin or eyes with running water for at least 15 minutes. Remove contaminated clothing and shoes. Keep patient at rest.
50 Oxidizer	**Fire:** May cause fire and react violently on contact with combustible. Reaction with fuels may be violent. **Explosion:** Mixture with fuels may explode.	**General:** No unnecessary personnel. Keep upwind. Identify and isolate hazard area. Self-contained breathing apparatus should be available. **Fire:** On small fires, use dry chemical or carbon dioxide. On large fires, use water spray or fog. Move ex-

Table 37-3. Summary of Potential Hazard and Immediate Action Information to be Shown in the HI Manual (cont'd).

Hazardous Information No. and Type	Potential Hazards	Immediate Action Information
50 (cont'd)	**Health:** Fire may produce poison gases.	posed containers from fire area, if without risk. Cool containers with water from maximum distance. Fight fire from maximum distance and from behind barrier. Use water spray to protect surrounding area. **Spill or Leak:** Avoid contact with spilled material. Stop leak if without risk. Keep spilled material away from combustibles. Dike large spills for later disposal. Collect small dry spills into clean, dry metal container and keep tightly covered. Use noncombustible absorbent material (sand) to collect small spills. Dilute liquid spill with large amounts of water. Dike for later disposal. **First Aid:** Remove contaminated clothing and shoes. Use standard first aid procedures.
51 Oxidizer, corrosive	**Fire:** May cause fire and reacts violently on contact with combustibles. Reactions with fuels may be violent. **Explosion:** Mixture with fuels may explode. **Health:** Contact with material may cause severe burns to skin and eyes. Fire may produce poison gas.	**General:** No unnecessary personnel. Keep upwind. Identify and isolate hazard area. Wear full-protective clothing. Self-contained breathing apparatus should be available. **Fire:** On small fires, use dry chemical or carbon dioxide. On large fires, use water spray or fog. Move exposed containers from fire area, if without risk. Cool containers with water from maximum distance. Fight fire from maximum distance and from behind barrier. Use water spray to protect surrounding area. **Spill or Leak:** Stop leak if without risk. Keep spilled material away from combustibles. Dike large spills for later disposal. Collect small, dry spills into clean, dry metal container and keep tightly covered. Use noncombustible absorbent material (sand) to collect small spills. Dilute liquid spill with large amounts of water. Dike for later disposal. **First Aid:** Remove to fresh air. Call physician. Remove contaminated clothing and shoes. In case of contact with material or water solution, immediately flush skin or eyes with running water for at least 15 minutes.
53 Oxidizer, poison, corrosive	**Fire:** May cause fire and reacts violently on contact with combustibles. Reaction with fuels may be violent. **Explosion:** Mixture with fuels may explode.	**General:** No unnecessary personnel. Keep upwind. Identify and isolate hazard area. Wear self-contained breathing apparatus and full-protective clothing. **Fire:** On small fires, use dry chemical or carbon dioxide. On large fires, use water spray or fog. Move exposed containers from fire area, if without risk. Cool

Table 37-3. Summary of Potential Hazard and Immediate Action Information to be Shown in the HI Manual (cont'd).

Hazardous Information No. and Type	Potential Hazards	Immediate Action Information
53 (cont'd)	**Health:** If inhaled, may be fatal. Contact with material may cause burns to skin and eyes. Runoff may pollute water supply.	containers with water from maximum distance. Fight fire from maximum distance and from behind barrier. Use water spray to protect surrounding area. **Spill or Leak:** Stop leak if without risk. Keep spilled material away from combustibles. Collect small solid spills into clean, dry metal container and keep tightly covered. Use noncombustible absorbent material (sand) to collect small spills. Dilute liquid spill with large amounts of water using spray or fog nozzle. Dike for later disposal. **First Aid:** Remove to fresh air. Call physician. If not breathing, give artificial respiration. If breathing is difficult, give oxygen. In case of contact with material or water solution, immediately flush skin or eyes with running water for at least 15 minutes. Keep patient at rest. Effects of contact or inhalation may be delayed.
54 Oxidizer, self-reactive, thermally unstable	**Fire:** May cause fire and react violently on contact with combustibles. Reaction with fuels may be violent. Heated container may rupture violently and produce flying missiles. **Explosion:** Mixture with fuels may explode. Decomposition with explosive violence may be caused by friction, shock, heat or contamination. Runoff to sewer may create explosion hazards. **Health:** Contact with material may cause severe burns to skin and eyes. Fire may produce poison gases.	**General:** No unnecessary personnel. Keep upwind. Identify and isolate hazard area. Wear full-protective clothing. Self-contained breathing apparatus should be available. **Fire:** On small fires, use dry chemical or carbon dioxide. On large fires, use flooding amounts of water. Cool containers with water from maximum distance. Withdraw from hazard area, if fire in cargo area is massive or advanced. If fire fighting is necessary, use unmanned hose holder or monitor from maximum distance or behind barrier. Use water spray to protect surrounding area. **Spill or Leak:** Stop leak if without risk. Keep spilled material away from combustibles. Use noncombustibles. Use noncombustible absorbent material (sand) to collect small spills. Dilute large liquid spills with large amounts of water, dike for later disposal. **First Aid:** Remove contaminated clothing and shoes. In case of contact with material and water solution, immediately flush skin or eyes with running water for at least 15 minutes. Use standard first aid procedures.
55 Oxidizer, corrosive, self-reactive, or thermally unstable	**Fire:** May cause fire and react violently on contact with combustibles. Reaction with fuels may be violent. Heated container may rupture violently and produce flying	**General:** No unnecessary personnel. Keep upwind. Identify and isolate hazard area. Wear full-protective clothing. Self-contained breathing apparatus should be available. **Fire:** On small fires, use dry chemical or carbon dioxide. On large fires, use flooding amounts of water. In

Table 37-3. Summary of Potential Hazard and Immediate Action Information to be Shown in the HI Manual (cont'd).

Hazardous Information No. and Type	Potential Hazards	Immediate Action Information
55 (cont'd)	missiles. Runoff to sewer may create fire or explosion hazard. **Explosion:** Mixture with fuels may explode. Decomposition with explosive violence may be caused by friction, shock, heat, or contamination. **Health:** Contact with material may cause severe burns to skin and eyes.	early stages, cool containers with water from maximum distance. Withdraw from hazard area, if fire in cargo area is massive or advanced. If fire fighting is necessary, use unmanned hose holder or monitor from maximum distance behind barrier. Use water spray to protect surrounding area. **Spill or Leak:** Stop leak if without risk. Keep spilled material away from combustibles. Use noncombustible absorbent material (sand) to collect small spills. Dilute liquid spill with large amounts of water. Dike for later disposal. **First Aid:** Remove to fresh air. Call physician. Remove contaminated clothing and shoes. In case of contact with material or water solution, immediately flush skin or eyes with running water for at least 15 minutes.
56 Oxidizer, poison, self-reactive or thermally unstable	**Fire:** May cause fire and reacts violently on contact with combustibles. Reaction with fuels may be violent. Heated container may rupture violently and produce flying missiles. **Explosion:** Mixture with fuels may explode. Decomposition with explosive violence may be caused by friction, shock, heat, or contamination. **Health:** Vapor, mist, or dust is poisonous, can be fatal if breathed in high concentrations. Contact with material may cause severe burns to skin and eyes. Runoff to sewer may create poison hazard.	**General:** No unnecessary personnel. Keep upwind. Identify and isolate hazard area. Wear self-contained breathing apparatus and full-protective clothing. **Fire:** On small fires, use dry chemical or carbon dioxide. On large fires, use flooding amounts of water. Cool containers with water from maximum distance. Withdraw from hazard area, if fire in cargo area is massive or advanced. If fire fighting is necessary, use unmanned hose or monitor from maximum distance behind barrier. Use water spray to protect surrounding area. **Spill or Leak:** Stop leak if without risk. Keep spilled material away from combustibles. Use water spray to reduce vapors. Use noncombustible absorbent material (sand) to collect small spills. Dilute liquid spill with large amounts of water. Dike for later disposal. **First Aid:** Remove to fresh air. Call physician. If not breathing, give artificial respiration. If breathing is difficult, give oxygen. Remove contaminated clothing and shoes. In case of contact with material or water solution, immediately flush skin and eyes with running water for at least 15 minutes. Keep patient at rest. Effects of contact or inhalation may be delayed.
57 Organic peroxide	**Fire:** May be ignited by heat, sparks, or open flame. May cause fire on contact with combustibles. Reaction with fuels may be violent. Heated container may rupture	**General:** No unnecessary personnel. Keep upwind. Identify and isolate hazard area. Wear full-protective clothing. Self-contained breathing apparatus should be available.

Table 37-3. Summary of Potential Hazard and Immediate Action Information to be Shown in the HI Manual (cont'd).

Hazardous Information No. and Type	Potential Hazards	Immediate Action Information
57 (cont'd)	violently and produce flying missiles even if water is applied for cooling. **Explosion:** Decomposition with explosive violence may be caused by friction, shock, heat, or contamination. **Health:** Contact with material may cause severe burns to skin and eyes.	**Fire:** In early stages, cool containers with water from maximum distance. On small fires, use dry chemical or carbon dioxide. On large fires, use flooding amounts of water. Withdraw from hazard area if fire in cargo area is massive or advanced. If fire fighting is necessary, use unmanned hose holder or monitor nozzles. **Spill or Leak:** Stop leak if without risk. Keep spilled material away from combustibles. Dilute large liquid spills with large amounts of water, dike for later disposal. **First Aid:** Remove to fresh air. Remove contaminated clothing and shoes. In case of contact with material or water solution, immediately flush skin or eyes with running water for at least 15 minutes. Keep patient at rest.
58 Organic peroxide, highly sensitive, needs refrigeration.	**Fire:** Autoignition may occur. May be ignited by heat, sparks, or open flames. Burns very rapidly and intensely, sometimes with flare-burning effect. May cause fire on contact with combustibles. Heated container may rupture violently and produce flying missiles. **Explosion:** Decomposition with explosive violence may be caused by loss of refrigeration, friction, shock, or contamination. **Health:** Contact with material may cause severe burns to skin and eyes.	**General:** No unnecessary personnel. Keep upwind. Identify and isolate hazard area. Wear full-protective clothing. Self-contained breathing apparatus should be available. **Loss of Cooling:** Specified temperature of material must be maintained to prevent explosion or dangerous decomposition. Obtain emergency source of cooling such as dry ice or liquid nitrogen. If no source of cooling can be obtained, evacuate area. **Fire:** In confined space such as refrigerator use liquid nitrogen or carbon dioxide if available to cool material and exclude air. Water will extinguish flames but may also raise material temperature above decomposition point. Withdraw from hazard area if fire in cargo area is massive or advanced. **Spill or Leak:** In case of small spill, deposit material in a shallow trench and ignite with torch from a safe distance. Flush area of spill with water. **First Aid:** Remove to fresh air. Call physician. Use standard first aid procedures.
59 Organic peroxide, extremely sensitive	**Fire:** Autoignition may occur. May be ignited by heat, sparks, or open flame. Burns very rapidly and intensely, sometimes with flare-burning effect. May cause fire on contact with combustibles. Reaction with fuels may be violent.	**General:** No unnecessary personnel. Keep upwind. Identify and isolate hazard area. Wear full-protective clothing. Self-contained breathing apparatus should be available. **Fire:** On small fires, use dry chemical or carbon dioxide. Withdraw from hazard area, if fire in cargo area

Table 37-3. Summary of Potential Hazard and Immediate Action Information to be Shown in the HI Manual (cont'd).

Hazardous Information No. and Type	Potential Hazards	Immediate Action Information
59 (cont'd)	Heated container may rupture violently and produce flying missiles. **Explosion:** Decomposition with explosive violence may be caused by friction, shock, heat, or contamination. **Health:** Contact with material may cause severe burns to skin and eyes.	is massive or advanced and use unmanned hose holder or monitor from maximum distance or behind barrier to protect surrounding area. **Spill or Leak:** Within hazard area: Eliminate ignition source. No flares, no smoking, no open flames. Stop leak if without risk. Keep spilled material away from combustibles. Flush area of spill with water. **First Aid:** Remove to fresh air. Call physician. In case of contact with material or water solution, immediately flush skin or eyes with running water for at least 15 minutes. Keep patient at rest.
60 Poison, highly toxic	**Fire:** Some material in this group cannot catch fire; others can catch fire but do not ignite readily. **Health:** Vapor, mist, or dust is poisonous, if breathed. Liquid or solid may cause death, if consumed. Contaminated water or material runoff may pollute water supply. Runoff to sewer may create poison hazard.	**General:** No unnecessary personnel. Keep upwind. Identify and isolate hazard area. Wear self-contained breathing apparatus and full-protective clothing. **Fire:** On small fires, use dry chemical or carbon dioxide. On large fires, use standard fire fighting agents. Move exposed containers from fire area, if without risk. **Spill or Leak:** Avoid contact with spilled material. Stop leak if without risk. Dike large spills for later disposal. Use water spray to reduce vapors. **First Aid:** Remove to fresh air. Call physician. If not breathing, give artificial respiration. If breathing is difficult, give oxygen. Remove contaminated clothing and shoes. In case of contact with material or water solution, immediately flush skin or eyes with running water for at least 15 minutes. Keep patient at rest. Effects of contact or inhalation may be delayed.
61 Poison, highly toxic, combustible	**Fire:** May be ignited by heat, sparks, or open flames. Heated container may rupture violently and produce flying missiles. Ignition of vapor may occur at some distance from leaking container. Vapor entering sewer or other closed spaces may create fire or explosion hazard. **Health:** Vapor, mist, or dust is poisonous, if breathed. Liquid or solid may cause death, if consumed. Contaminated water or material runoff may pollute water supply. Runoff to sewer may create poison hazard.	**General:** No unnecessary personnel. Keep upwind. Identify and isolate hazard area. Wear self-contained breathing apparatus and full-protective clothing. **Fire:** On small fires, use dry chemical or carbon dioxide. On large fires, use standard fire fighting agents. Do not approach ends of horizontal tanks. Move exposed containers from fire area, if without risk. Cool containers with water. Continue cooling after fires have been extinguished.

Table 37-3. Summary of Potential Hazard and Immediate Action Information to be Shown in the HI Manual (cont'd).

Hazardous Information No. and Type	Potential Hazards	Immediate Action Information
61 (cont'd)		**Spill or Leak:** Within hazard area: Eliminate ignition source. No flares, no smoking, no open flames. Stop leak if without risk. Dike large spills for later disposal. Use water spray to reduce vapors. Use noncombustible absorbent material (sand) to collect small spills. **First Aid:** Remove to fresh air. Call physician. If not breathing, give artificial respiration. If breathing is difficult, give oxygen. Remove contaminated clothing and shoes. In case of contact with material or water solution, immediately flush skin or eyes with running water for at least 15 minutes. Keep patient at rest. Effects of contact or inhalation may be delayed.
62 Poison, extremely toxic	**Fire:** Some material in this group cannot catch fire; others can catch fire but do not ignite readily. **Health:** Vapor, mist, or dust is poisonous, can be fatal if breathed in high concentrations. Contact with material may cause severe burns to skin and eyes. Small amounts of liquid or solid may cause death, if consumed. Contaminated water or material runoff may pollute water supply. Runoff to sewer may create poison hazard.	**General:** No unnecessary personnel. Keep upwind. Identify and isolate hazard area. Wear self-contained breathing apparatus and full-protective clothing. **Fire:** On small fires, use dry chemical or carbon dioxide. On large fires, use standard fire fighting agents. Move exposed containers from fire area, if without risk. **Spill or Leak:** Avoid contact with spilled material. Stop leak if without risk. Dike large spills for later disposal. Use water spray to reduce vapors. **First Aid:** Remove to fresh air. Call physician. If not breathing, give artificial respiration. If breathing is difficult, give oxygen. Remove contaminated clothing and shoes. In case of contact with material or water solution, immediately flush skin or eyes with running water for at least 15 minutes. Keep patient at rest. Effects of contact or inhalation may be delayed.
64 Poison, poisonous through skin absorption, extremely or highly toxic	**Fire:** Some material in this group cannot catch fire; others can catch fire but do not ignite readily. **Health:** Vapor, mist, or dust is poisonous, can be fatal if breathed in high concentrations. Poisonous by skin absorption. Contact with material may cause severe burns to skin and eyes. Small amounts of liquid or solid may cause death, if consumed. Contaminated water or material runoff may pollute water supply. Runoff to sewer may create poison hazard.	**General:** No unnecessary personnel. Keep upwind. Identify and isolate hazard area. Wear self-contained breathing apparatus and full-protective clothing. **Fire:** On small fires, use dry chemical or carbon dioxide. On large fires, use standard fire fighting agents. Move exposed containers from fire area, if without risk. **Spill or Leak:** Do not come into contact with spilled material. Stop leak if without risk. Dike large spills for later disposal. **First Aid:** Remove to fresh air. Call physician. If not breathing, give artificial respiration. If breathing is difficult, give oxygen. Remove contaminated clothing and shoes. Speed in removing liquid from

Table 37-3. Summary of Potential Hazard and Immediate Action Information to be Shown in the HI Manual (cont'd).

Hazardous Information No. and Type	Potential Hazards	Immediate Action Information
64 (cont'd)		skin is of extreme importance. In case of contact with material or water solution, immediately flush skin or eyes with running water for at least 15 minutes. Keep patient at rest. Effects of contact or inhalation may be delayed.
65 Poison, extremely toxic, flammable	**Fire:** May be ignited by heat, sparks, or open flames. Heated container may rupture violently and produce flying missiles. Ignition of vapor may occur at some distance from leaking container. Vapor entering sewer or other closed spaces may create fire or explosion hazard. **Health:** Vapor, mist, or dust is poisonous, can be fatal if breathed in high concentrations. Contact with material may cause severe burns to skin and eyes. Small amounts of liquid or solid may cause death, if consumed. Contaminated water or material runoff may pollute water supply. Runoff to sewer may create poison hazard.	**General:** No unnecessary personnel. Keep upwind. Identify and isolate hazard area. Wear self-contained breathing apparatus and full-protective clothing. **Fire:** On small fires, use dry chemical or carbon dioxide. On large fires, use standard fire fighting agents. Do not approach ends of horizontal tanks. Move exposed containers from fire area, if without risk. Cool containers with water. Continue cooling after fires have been extinguished. **Spill or Leak:** Within hazard area: Eliminate ignition source. No flares, no smoking, no open flames. Avoid contact with spilled material. Stop leak if without risk. Dike large spills for later disposal. Use water spray to reduce vapors. Use noncombustible absorbent material, (sand) to collect small spills. **First Aid:** Remove to fresh air. Call physician. If not breathing, give artificial respiration. If breathing is difficult, give oxygen. Remove contaminated clothing and shoes. In case of contact with material or water solution, immediately flush skin or eyes with running water for at least 15 minutes. Keep patient at rest. Effects of contact or inhalation may be delayed.
67 Poison, flammable, poisonous through skin absorption, extremely toxic	**Fire:** May be ignited by heat, sparks, or open flames. Heated container may rupture violently and produce flying missiles. Ignition of vapor may occur at some distance from leaking container. Vapor entering sewer or other closed space may create fire or explosion hazard. **Health:** Vapor, mist, or dust is poisonous, can be fatal if breathed in high concentrations. Contact with material may cause severe burns to skin and eyes. Poisonous by skin absorption. Small amounts of liquid or solid may cause death, if consumed. Contaminated water or material runoff may pollute	**General:** No unnecessary personnel. Keep upwind. Identify and isolate hazard area. Wear self-contained breathing apparatus and full-protective clothing. **Fire:** On small fires, use dry chemical or carbon dioxide. On large fires, use standard fire fighting agents. Do not approach ends of horizontal tanks. Move exposed containers from fire area, if without risk. Cool containers with water. Continue cooling after fires have been extinguished. **Spill or Leak:** Within hazard area: Eliminate ignition source. No flares, no smoking, no open flames. Do not come into contact with spilled material. Stop leak if without risk. Dike large spills for later disposal. Use water spray to reduce vapors. Use noncombustible absorbent material, (sand) to collect small spills.

Table 37-3. Summary of Potential Hazard and Immediate Action Information to be Shown in the HI Manual (cont'd).

Hazardous Information No. and Type	Potential Hazards	Immediate Action Information
67 (cont'd)	water supply. Runoff to sewer may create poison hazard.	**First Aid:** Remove to fresh air. Call physician. If not breathing, give artificial respiration. If breathing is difficult, give oxygen. Remove contaminated clothing and shoes. Speed in removing liquid from skin is of extreme importance. In case of contact with material or water solution, immediately flush skin or eyes with running water for at least 15 minutes. Keep patient at rest. Effects of contact or inhalation may be delayed.
70 Radioactive, low hazard	**Health:** Degree of hazard due to radioactivity will vary depending upon type, quantity, and form of the material. Hazard may be from *internal* radiation from breathing gases, vapor, or dust from airborne material or contamination of skin, open cuts, sores; it may be from *external* radiation (as from X-rays) from contamination on skin or from exposure to unshielded radioactive material. Radiation hazard is generally of lower order and does not pose an immediate threat to life or health.	**General:** No unnecessary personnel. Keep upwind. Identify and isolate hazard area. Wear full-protective clothing. Self-contained breathing apparatus should be available. **Fire:** Use standard fire fighting agents. Avoid contact with leaking or damaged packages. Move undamaged packages out of fire zone if without risk. Delay cleanup until arrival of qualified radiation monitoring assistance. **Spill or Leak:** Avoid contact with leaking or damaged packages. If damaged packages must be moved, use gloves to place damaged packages in metal containers if available. Shut off liquid leak or dike flow or use absorbent materials to contain leakage. Move undamaged packages from spill area. Delay cleanup until arrival of qualified radiation monitoring assistance. **First Aid:** Call physician. Use standard first aid procedures. Assume radioactive contamination on persons and equipment close to damaged packages or spilled material. Remove contaminated clothing and wash or shower with soap and water, if possible. Advise rescue personnel and physicians that injured persons may be radioactively contaminated.
71 Radioactive	**Health:** Degree of hazard due to radioactivity will vary depending upon type, quantity, and form of the material. Hazard may be from *internal* radiation from breathing gases, vapors, or dust from airborne material or contamination of skin, open cuts, sores. It may be from external radiation (as from X-rays) from contamination on skin or from exposure to unshielded radioactive material. Prolonged exposure may be threat to health or life.	**General:** No unnecessary personnel. Keep upwind. Identify and isolate hazard area. Wear self-contained breathing apparatus and full-protective clothing. **Fire:** Use standard fire fighting agents. Fight fire from maximum distance. Avoid contact with leaking or damaged packages. Do not move damaged packages. Move undamaged packages out of fire zone if without risk. Delay cleanup until arrival of qualified radiation monitoring assistance. **Spill or Leak:** Avoid contact with leaking or damaged packages. Prevent spread of loose material by diking or other suitable means. Limit entries to

Table 37-3. Summary of Potential Hazard and Immediate Action Information to be Shown in the HI Manual (cont'd).

Hazardous Infor- mation No. and Type	Potential Hazards	Immediate Action Information
71 (cont'd)		hazard area to shortest possible time. Alternate persons used for entries if possible. Do not enter hazard area unless necessary to rescue injured persons or retard flow of material from massive spills or leaks. Delay cleanup until arrival of qualified radiation monitoring assistance.
		First Aid: Call physician. Use standard first aid procedures. Assume radioactive contamination on persons or equipment close to damaged packages or spilled material. Remove contaminated clothing and wash or shower with soap and water, if possible. Observable effects of serious radiation exposure may be delayed. Advise rescue personnel and physicians that injured persons may be radioactively contaminated.
72 Radioactive, oxidizer	**Fire:** May cause fire on contact with combustibles. Reaction with fuels may be violent. **Health:** Material is of relatively low order of hazard with regard to external radiation (as from X-rays). Primary radiation hazard is internal, caused by breathing gases, vapors, dust from airborne material or from contamination of skin, open cuts, sores. Fire may produce poisonous gases.	**General:** No unnecessary personnel. Keep upwind. Identify and isolate hazard area. Wear self-contained breathing apparatus and full-protective clothing. **Fire:** On small fires, use dry chemical or carbon dioxide. On large fires, use flooding amounts of water. Move exposed packages from hazard area. Large volumes of burning material may cause scattering of molten material when water is applied. Delay cleanup until arrival of qualified radiation monitoring assistance. **Spill or Leak:** Gather spilled material and place in closed metal containers as soon as possible to prevent combustion. Move undamaged packages away from spill area. Delay cleanup until arrival of qualified radiation monitoring assistance. **First Aid:** Call physician. Use standard first aid procedures. Assume radioactive contamination on persons or equipment close to damaged packages or spilled material. Wash hands and exposed parts of the body with soap and water, and shower, if possible. Advise rescue personnel and physicians of possibility of radioactive contamination.
73 Radioactive, corrosive	**Health:** Material is of a relatively low order of hazard with regard to external radiation (as from X-rays). Primary radiation hazard is internal, caused by breathing gases or vapor or from contamination of skin, open cuts, sores. Vapor, dust or mist is poisonous, can be fatal if breathed in high concentrations. Contact with material may cause severe	**General:** No unnecessary personnel. Keep upwind. Identify and isolate hazard area. Wear self-contained breathing apparatus and full-protective clothing. **Fire:** On small fires, use dry chemical or carbon dioxide. On large fires, use flooding amounts of water. Cool containers with water from maximum distance. Delay cleanup until arrival of qualified radiation monitoring assistance.

Table 37-3. Summary of Potential Hazard and Immediate Action Information to be Shown in the HI Manual (cont'd).

Hazardous Information No. and Type	Potential Hazards	Immediate Action Information
73 (cont'd)	burns to skin and eyes. Fire may produce poisoning gases.	**Spill or Leak:** Stop leak if without risk. Dilute spill with large amounts of water. Dike for later disposal. Delay cleanup until arrival of qualified radiation monitoring assistance. **First Aid:** Call physician. Use standard first aid procedures. Assume radioactive contamination on persons or equipment close to spill area. Wash all exposed parts of body with soap and water, and shower if possible. Advise personnel and physicians of possibility of radioactive contamination.
74 Radioactive, pyrophoric	**Fire:** May catch fire spontaneously in air. Burns very rapidly and intensely, sometimes with flare-burning effect. May reignite after fire is extinguished. **Health:** Material is of relatively low order of hazard with regard to external radiation (as from X-rays). Primary radiation hazard is internal, caused by breathing vapors or dusts from airborne material or by contamination of skin, open cuts, sores.	**General:** No unnecessary personnel. Keep upwind. Identify and isolate hazard area. Wear self-contained breathing apparatus and full-protective clothing. **Fire:** On small fires, use dry chemical or sand. If fire cannot be controlled, use flooding amounts of water. Delay cleanup until arrival of qualified radiation monitoring assistance. **Spill or Leak:** Gather spilled material using shovel and place under water or mineral oil in metal container as soon as possible to prevent self-ignition. Move undamaged packages from spill area. Delay cleanup until arrival of qualified radiation monitoring assistance. **First Aid:** Call physician. Use standard first aid procedures. Assume radioactive contamination on persons or equipment close to damaged packages or spilled material. All exposed persons should wash hands and exposed parts of body with soap and water, and shower, if possible. Advise rescue personnel and physicians of possibility of radioactive contamination.
78 Radioactive poison, corrosive, for use in uranium hexafluoride (UF_6) only	**Health:** Material is of a relatively low order of hazard with regard to external radiation (as from X-rays). Vapor, dust or mist is poisonous, can be fatal if breathed in high concentrations. Contact with material may cause severe burns to skin and eyes. The reaction product with air is readily visible as a white cloud, settling as a dust on surfaces.	**General:** No unnecessary personnel. Keep upwind. Identify and isolate hazard area. Wear self-contained breathing apparatus and full-protective clothing. **Fire:** On small fires, use dry chemical or carbon dioxide. On large fires, use water spray or fog. Move exposed containers from fire area, if without risk. Keep undamaged packages cool with large volumes of water. Delay cleanup until arrival of qualified radiation monitoring assistance. **Spill or Leak:** Attempt to plug releases from container openings using wooden plugs or freeze leakage by cooling with water stream at point of opening. Use water spray to reduce vapors. Pressurized CO_2 may also be effective in "freezing" leakage. Dilute

Table 37-3. Summary of Potential Hazard and Immediate Action Information to be Shown in the HI Manual (cont'd).

Hazardous Information No. and Type	Potential Hazards	Immediate Action Information
78 (cont'd)		spill with large amounts of water, dike for later disposal. Delay cleanup until arrival of qualified radiation monitoring assistance.
		First Aid: Call physician. Use standard first aid procedures. Assume radioactive contamination on persons or equipment close to damaged packages in spill area. All exposed persons should wash hands and exposed parts of the body with soap and water, and shower, if possible. Advise rescue personnel and physicians of possibility of radioactive contamination and chemical burns from exposure to spilled material.
79 Radioactive for use on acid solutions of plutonium nitrate	**Health:** Direct external radiation (as from X-rays) is relatively low. Spilled material is extremely hazardous with regard to internal radiation from contact with skin, cuts, wounds, or from breathing airborne dusts and fumes. An extremely radiotoxic material when taken into the body.	**General:** No unnecessary personnel. Keep upwind. Identify and isolate hazard area. Wear self-contained breathing apparatus and full-protective clothing.
		Fire: Use standard fire fighting agents. Fight fire from maximum distance. **Do not** come into contact with leaking or **damaged** packages. **Do not** move **damaged** packages. Move **undamaged** packages from fire zone, if without risk. Delay cleanup until arrival of qualified radiation monitoring assistance.
		Spill or Leak: Do not come in contact with leaking or damaged packages or enter spill area unless absolutely necessary to save life. Delay any actions until arrival of qualified radiation monitoring assistance unless action is needed to rescue injured persons.
		First Aid: Call physician. Use standard first aid procedures. Assume highly radioactive contamination on persons or equipment close to damaged packages or spilled material. Remove contaminated clothing and shower thoroughly with soap and water. Observable effects of serious inhalation or absorption of spilled material may be delayed. Advise rescue personnel and physicians that injured persons, clothing, etc., may be contaminated with highly radioactive material (plutonium).
80 Corrosive	**Fire:** Some material in this group cannot catch fire; others can catch fire but do not ignite readily.	**General:** No unnecessary personnel. Keep upwind. Identify and isolate hazard area. Wear full-protective clothing.
	Health: Vapor may be irritating, if breathed. Contact with material may cause severe burns to skin and eyes. Contaminated water or material runoff may pollute water supply.	**Fire:** On small fires, use dry chemical or carbon dioxide. On large fires, use standard fire fighting agents. Move exposed containers from fire area, if without risk. Cool containers with water.
		Spill or Leak: Stop leak if without risk. Dilute spill with large amounts of water. Dike for later disposal.

Table 37-3. Summary of Potential Hazard and Immediate Action Information to be Shown in the HI Manual (cont'd).

Hazardous Information No. and Type	Potential Hazards	Immediate Action Information
80 (cont'd)		**First Aid:** Remove to fresh air. Call physician. In case of contact with material or water solution, immediately flush skin or eyes with running water for at least 15 minutes. Remove contaminated clothing and shoes. Keep patient at rest. Effects of contact or inhalation may be delayed.
81 Corrosive, poison	**Fire:** Some material in this group cannot catch fire; others can catch fire but do not ignite readily. **Health:** Vapor, mist, or dust is poisonous if breathed. Contact with material may cause severe burns to skin and eyes. Contaminated water or material runoff may pollute water supply. Runoff to sewer may create poison hazard.	**General:** No unnecessary personnel. Keep upwind. Identify and isolate hazard area. Wear self-contained breathing apparatus and full-protective clothing. **Fire:** On small fires, use dry chemical or carbon dioxide. On large fires, use standard fire fighting agents. Move exposed containers from fire area, if without risk. Cool containers with water. **Spill or Leak:** Stop leak if without risk. Dilute spill with large amounts of water. Dike for later disposal. **First Aid:** Remove to fresh air. Call physician. In case of contact with material or water solution, immediately flush skin or eyes with running water for at least 15 minutes. Remove contaminated clothing and shoes. Keep patient at rest. Effects of contact or inhalation may be delayed.
82 Corrosive, heat of dilution	**Fire:** Some material in this group cannot catch fire; others can catch fire but do not ignite readily. **Explosion:** Explosive concentrations of gas may accumulate in tanks containing acid. **Health:** Vapor may be irritating, if breathed. Contact with material may cause severe burns to skin and eyes. Contaminated water or material runoff may pollute water supply.	**General:** No unnecessary personnel. Keep upwind. Identify and isolate hazard area. Wear full-protective clothing. Self-contained breathing apparatus should be available. **Fire:** On small fires, use dry chemical or carbon dioxide. On large fires, use water spray or fog. Move exposed containers from fire area, if without risk. Do not get water inside containers. **Spill or Leak:** Stop leak if without risk. Dilute spill with large amounts of water. Dike for later disposal. Do not get water inside containers. **First Aid:** Remove to fresh air. Call physician. In case of contact with material or water solution, immediately flush skin or eyes with running water for at least 15 minutes. Remove contaminated clothing and shoes. Keep patient at rest. Effects of contact or inhalation may be delayed.
83 Corrosive, combustible	**Fire:** May be ignited by heat, sparks, or open flames. Heated container may rupture violently and produce flying missiles. Ignition of vapor may occur at some distance from leaking container. Vapor entering sewer or other	**General:** No unnecessary personnel. Keep upwind. Identify and isolate hazard area. Wear full-protective clothing. Self-contained breathing apparatus should be available. **Fire:** On small fires, use dry chemical or carbon dioxide. On large fires, use standard fire fighting agents. Do not approach ends of horizontal tanks. Move

Table 37-3. Summary of Potential Hazard and Immediate Action Information to be Shown in the HI Manual (cont'd).

Hazard Infor- mation No. and Type	Potential Hazards	Immediate Action Information
83 (cont'd)	closed spaces may create fire or explosion hazard. **Health:** Vapor may be irritating, if breathed. Contact with material may cause severe burns to skin and eyes. Contaminated water or material runoff may pollute water supply. Runoff to sewer may create poison hazard.	exposed containers from fire area, if without risk. Cool containers with water. Continue cooling after fires have been extinguished. **Spill or Leak:** Within hazard area: Eliminate ignition source. No flares, no smoking, no open flames. Stop leak if without risk. Use water spray to reduce vapors. Dike large spills for later disposal. Use noncombustible absorbent material (sand) to collect small spills. **First Aid:** Remove to fresh air. Call physician. In case of contact with material or water solution, immediately flush skin or eyes with running water for at least 15 minutes. Remove contaminated clothing and shoes. Keep patient at rest. Effects of contact or inhalation may be delayed.
84 Corrosive, poison, heat of dilution	**Fire:** Some material in this group cannot catch fire; others can catch fire but do not ignite readily. **Explosion:** Explosive concentrations of gas may accumulate in tanks containing acids. **Health:** Vapor, mist, or dust is poisonous, if breathed. Contact with material may cause severe burns to skin and eyes. Contaminated water or material runoff may pollute water supply. Runoff to sewer may create poison hazard.	**General:** No unnecessary personnel. Keep upwind. Identify and isolate hazard area. Wear self-contained breathing apparatus and full protective clothing. **Fire:** On small fires, use dry chemical or carbon dioxide. On large fires, use water spray or fog. Move exposed containers from fire area, if without risk. Do not get water inside containers. **Spill or Leak:** Stop leak if without risk. Dilute spill with large amounts of water. Dike for later disposal. Do not get water inside containers. **First Aid:** Remove to fresh air. Call physician. In case of contact with material or water solution, immediately flush skin or eyes with running water for at least 15 minutes. Remove contaminated clothing and shoes. Keep patient at rest. Effects of contact or inhalation may be delayed.
85 Corrosive, combustible, poison	**Fire:** May be ignited by heat, sparks, or open flames. Heated container may rupture violently and produce flying missiles. Ignition of vapor may occur at some distance from leaking container. Vapor entering sewer or other closed spaces may create fire or explosion hazard. **Health:** Vapor, mist, or dust is poisonous, if breathed. Contact with material may cause severe burns to skin and eyes. Contaminated water or material runoff may	**General:** No unnecessary personnel. Keep upwind. Identify and isolate hazard area. Wear self-contained breathing apparatus and full protective clothing. **Fire:** On small fires, use dry chemical or carbon dioxide. On large fires, use standard fire fighting agents. Do not approach ends of horizontal tanks. Move exposed containers from fire area, if without risk. Cool containers with water. Continue cooling after fires have been extinguished. **Spill or Leak:** Within hazard area: Eliminate ignition source. No flares, no smoking, no open flames. Avoid contact with spilled material. Stop leak if without risk. Dilute spill with large amounts of water. Dike for later disposal.

Table 37-3. Summary of Potential Hazard and Immediate Action Information to be Shown in the HI Manual (cont'd).

Hazardous Information No. and Type	Potential Hazards	Immediate Action Information
85 (cont'd)	may pollute water supply. Runoff to sewer may create poison hazard.	**First Aid:** Remove to fresh air. Call physician. In case of contact with material or water solution, immediately flush skin or eyes with running water for at least 15 minutes. Remove contaminated clothing and shoes. Keep patient at rest. Effects of contact or inhalation may be delayed.
86 Corrosive, combustible, heat of dilution	**Fire:** May be ignited by heat, sparks, or open flames. Heated container may rupture violently and produce flying missiles. Ignition of vapor may occur at some distance from leaking container. Vapor entering sewer or other closed spaces may create fire or explosion hazard. **Explosion:** Explosive concentrations of gas may accumulate in tanks containing acid. **Health:** Vapor or mist may be irritating, if breathed. Contact with material may cause severe burns to skin and eyes. Contaminated water or material runoff may pollute water supply. Runoff to sewer may create poison hazard.	**General:** No unnecessary personnel. Keep upwind. Identify and isolate hazard area. Wear full-protective clothing. Self-contained breathing apparatus should be available. **Fire:** On small fires, use dry chemical or carbon dioxide. On large fires, use water spray or fog. Move exposed containers from fire area, if without risk. Cool containers with water. Do not get water inside containers. **Spill or Leak:** Within hazard area: Eliminate ignition source. No flares, no smoking, no open flames. Stop leak if without risk. Dilute spill with large amounts of water. Dike for later disposal. Do not get water inside containers. **First Aid:** Remove to fresh air. Call physician. In case of contact with material or water solution, immediately flush skin or eyes with running water for at least 15 minutes. Remove contaminated clothing and shoes. Keep patient at rest. Effects of contact or inhalation may be delayed.
87 Corrosive, combustible, poison, heat of dilution	**Fire:** May be ignited by heat, sparks, or open flames. Heated container may rupture violently and produce flying missiles. Ignition of vapor may occur at some distance from leaking container. Vapor entering sewer or other closed spaces may create fire or explosion hazard. **Explosion:** Explosive concentrations of gas may accumulate in tanks containing acid. **Health:** Vapor or mist is poisonous if breathed. Contact with material may cause severe burns to skin and eyes. Contaminated water or material runoff may pollute water supply. Runoff to sewer may create poison hazard.	**General:** No unnecessary personnel. Keep upwind. Identify and isolate hazard area. Wear self-contained breathing apparatus and full-protective clothing. **Fire:** On small fires, use dry chemical or carbon dioxide. On large fires, use water spray or fog. Move exposed containers from fire area, if without risk. Do not get water inside containers. **Spill or Leak:** Within hazard area: Eliminate ignition source. No flares, no smoking, no open flames. Stop leak if without risk. Dilute spill with large amounts of water. Dike for later disposal. Do not get water inside container. **First Aid:** Remove to fresh air. Call physician. In case of contact with material or water solution, immediately flush skin or eyes with running water for at least 15 minutes. Remove contaminated clothing and shoes. Keep patient at rest. Effects of contact or inhalation may be delayed.

Table 37-3. Summary of Potential Hazard and Immediate Action Information to be Shown in the HI Manual (cont'd).

Class of Hazard	Signal Word	Statement of Hazard
Ingestion		
Highly toxic	Danger!	May be fatal if swallowed.
Toxic	Warning!	Harmful if swallowed.
Absorption		
Highly toxic	Danger!	May be fatal if absorbed through skin.
Toxic	Warning!	Harmful if absorbed through skin.
Inhalation		
Highly toxic	Danger!	May be fatal if inhaled.
Toxic	Warning!	Harmful if inhaled.
Strong sensitizer, lungs	Warning!	May cause allergic respiratory reaction.
Physiological inert	Caution!	Reduces oxygen available for breathing.
Contact		
Corrosive, eye	Danger!	Causes (severe) eye burns.
Corrosive, eye and skin	Danger!	Causes (severe) burns.
Irritant, eye	Warning!	Causes eye irritation.
Irritant, eye and skin	Warning!	Causes irritation.
Strong sensitizer, skin	Warning!	May cause allergic skin reaction.
Flammability		
Extremely flammable liquid	Danger!	Extremely flammable.
Flammable liquid	Warning!	Flammable.
Flammable solid	Warning!	Flammable.
Flammable gas	Danger!	Extremely flammable under pressure.
Combustible liquid	Caution!	Combustible.
Pyrophoric substance	Danger!	Extremely flammable, catches fire if exposed to air.
Strong oxidizer	Danger!	Strong oxidizer — contact with other material may cause fire.

Table 37-4. Manufacturing Chemists Association Precautionary Statements for Hazardous Chemical Labels.

Acknowledgments

The authors wish to thank the many people who made this text possible.

- Deputy Fire Chief Dwight Lechner and all the members of the Everett (WA) Fire Department.

- Jerry Brockey and Robert Logue of South Seattle (WA) Community College.

- Members of the East Glenville (NY) Fire Department.

- Fire science faculty and students of the Schenectady (NY) County Community College.

CONTRIBUTORS

- Houston (TX) Fire Department; Gary Morgan, photographer: section 1 and section 8 photographs.

- Los Angeles (CA) City Fire Department: section 4 photograph.

- Seattle (WA) Fire Department; John Philbin, Assistant Chief: title page, section 2, section 6, and section 10 photographs.

- Scotia (NY) Fire Department; H.L. Woodworth, photographer: section 9 photograph.

- *Telegraph,* Alton (IL): section 5 photograph.

- University of Maryland: section 7 photograph.

- *Schenectady Gazette,* Schenectady (NY); Sid Brown, photographer: section 3 (bottom) photograph.

- Robert P. Morris: section 3 (top) photograph.

- National Fire Protection Association.

- U.S. Department of Transportation.

DELMAR STAFF

Publications Director — Alan N. Knofla

Editorial Consultant — Donald Favreau

Source Editor — Elinor Gunnerson

Copy Editor — Judith Barrow

Director of Manufacturing and Production — Frederick Sharer

Production Specialists — Betty Michelfelder, Jean LeMorta, Patti Manuli, Debbie Monty, Alice Schielke, Lee St. Onge

Illustrators — Anthony Canabush, George Dowse, Michael Kokernak

Selected portions of the manuscript were classroom tested by Leroy Schieler at South Seattle Community College and by Denis Pauzé at Schenectady County Community College.

Index

A

ABS, 113
Acetic acid, 151, 185
Acetylene, 62-63
    fire situations involving, 63
    hazards connected with storage, 63
    storing, 62-63
Acetylene series, 95
Acetyl peroxide, 135-36
ACGIH accepted values, 161, 181
Acids, 105
    acetic acid, 151
    characteristics, 147-48
    formic acid, 151
    halogen acid, 149-50
    hydrofluoric, 149
    nitrating acid, 150
    nitric acid, 150
    organic acid, 151
    perchloric acid, 150-51
    phenol, 151
    sulfuric acid, 149
Acrolein, 167
Acrylics, 113
Alcohols, 100-102
    fire fighting procedures, 101-102
    glycols, 101
Aldehydes, 104-105
Alkali metals
    fire fighting procedures, 81-82
    use, areas of, 80-81
Alkaline earth metals, 82-84
Allotropes, 88
Alpha particles, 196
Aluminum, 84
Amides, 105-106
Amines, 102-103, 153
Ammonia, 185
Ammonia dynamite, 130
Ammonia gas
    leaks, handling of, 70
    shipping methods, 69-70
    storage, 69
Ammonium nitrate, 121
Anesthetics, 160
Aniline, 155
Anoxemia, 169
Anoxia, 164
Aqua fortis. *See* Nitric acid
Aramid, 114
Aromatic hydrocarbons, 95-96
Asphyxiants, 159
Atomic number, 1
Atomic weight, 1-2
Atoms
    combining capacity of, 8
    electronic configuration, 2-3
    nuclear atom, 1-2
Azides, 121, 132

B

Balanced chemical equation, 9
Barium, 82
Bases
    amines, 153
    ammonium hydroxide, 154-55
    aniline, 155
    caustic potash, 154
    diethylamine, 156
    hazards of, 155-56
    sodium hydroxide, 153-54
Battery acid. *See* Sulfuric acid
Benzene, 95
Beryllium, 82
Beryllosis, 169
Beta rays, 196
Binary compounds, 8
Black phosphorus, 89
Black powder, 131
BLEVE, 67
Bond energies, 15
Boranes, 143
Boron, 84
Boyle's Law, 22
Brisant explosives, 29
Brissance, 126
Bronchitis, 169
Butadiene, 115
Butane, 65-66
Butyl alcohol, 101

C

Calcium, 82
Carbides, 143-44
Carbon, 88
Carbon dioxide, 4
Carbon monoxide, 4, 167
Carbonates, 153
Carbon tetrachloride, 51
Cargo tanks, 58
Catalyst, 112
Caustic soda. *See* Sodium hydroxide
Celluloid, 113
Cellulose, 113
Cellulose acetate, 113
Cellulose nitrate, 113
Charles' Law, 22
Chemical asphyxiants, 159
Chemical change, laws of, 3-4
Chemical Data Guide for Bulk Shipment by Water, 205
Chemical equations, 9
Chemical extinguishing agents, 49-52
Chemical formulas, 8-9
Chemical inertness, 5
Chemical reactions, 4-5
Chlorine, 166
Chlorine gas, 68-69
    controlling fires near, 69
    fire hazards of, 68-69
    leaks, handling of, 69
    shipping, 69
Chlorpicrin, 166
C, H, O, N. systems, 32

CHO system, combustion of, 31
Class A fires, 49
Class B fires, 49
Class C fires, 49
Class D fires, 49-50
Closed shell structures, 4
Coefficient, 9
Collodion, 131
Combustible, 39
Combustible materials
    metals, 79-85
    nonmetals, 88-91
    plastic, 109-17
    substituted hydrocarbons, 100-102
Combustion
    chemical extinguishing agent,
        reactivity with, 49-52
    flammability limits, 35
    flash point, 36-39
    ignition temperature, 39
    oxygen, reactivity with, 42-43
    specific gravity, 40
    vapor density, 40
    vs. explosion, 25-27
    water, reactivity with, 45-47
    water solubility and, 39-40
Combustion behavior, 35-40
Combustion mechanism, 11-13
Compounds, 3
    naming, 8-9
Compressed gases
    acetylene, 62-63
    ammonia gas, 69-70
    chlorine gas, 68-69
    defined, 55
    liquefied, 64-68
    MAPP gas, 63-64
    natural gas, 60-61
    oxygen, 59-60
Conjunctivitis, 169
Conservation of mass, law of, 4, 9
Constant composition, law of, 4
Corrosive materials
    acids, 147-51
    bases, 153-56
Corrosive poisons, 184-85
    acetic acid, 185
    ammonia, 185
    nitric acid, 185
    phosphoric acid, 185
    sulfuric acid, 185
Critical temperature, 65
Cryogenic gases
    defined, 72
    liquid argon, 75
    liquid fluorine, 74
    liquid helium, 75
    liquid krypton, 75
    liquid natural gas, 75-76
    liquid neon, 75
    liquid nitrogen, 74-75

Cryogenic gases (cont'd)
   liquid oxygen, 72-74
   liquid xenon, 75
Cyanides, 166-67
Cyanosis, 169

D

Deflagrate, 29
Delay igniters, 122
Detonating cord, 132
Detonating devices, 122
Detonation cap, 122
Detonators, 132
Deuterium, 2
Diatomic molecule, 4
Diethylamine, 156
Dioxides, 165
Dipping acid. *See* Sulfuric acid
Dip tube, 69
Dry chemical extinguishers, 11
Dynamite
   ammonia, 130
   gelatin, 130-31
   straight, 130
Dyspnea, 169

E

Edema, 169
Electron, 1
Electrostatic attraction, 1
Elements, 2
Endothermic reactions, 14-15
Epoxies, 113
Esters, 105
Ethers, 106
Ethyl alcohol, 100-101
Ethylene, 115
Ethyl silicate, 165
Exothermic reactions, 14-15
Explosion mechanisms, 25-27
Explosions, 29-30
Explosive detonators, 132
Explosive materials
   azides, 132
   black powder, 131
   detonating devices for, 122
   dynamite, 130-31
   fulminates, 132
   nitric ester explosives, 127-28
   nitroexplosives, 124-26
   pyrotechnics, 131-32
Explosives, principle of
   shock sensitivity, 122
   thermal sensitivity, 121-22
Exposure
   acute, 32
   chronic, 32
   local, 32
   subacute, 32
   systemic, 32
Extinguishing agents
   carbon tetrachloride, 51
   halocarbons, 51-52

Extinguishing agents (cont'd)
   halogens, 51
   properties of, 50-52

F

Fibrosis, 169
Fire extinguishers
   fire classification and, 49-50
   fire size and, 50
Fire Protection Guide on Hazardous
   Materials, 205
Fires, classification, 49-50
Flammability limits, 35
Flammable liquid, 39, 97
Flash point, 35, 36-39
Fluoroplastics, 114
Formic acid, 151
Frangible disc, 56
Free radicals, 12
Fulminates, 121
   mercury fulminate, 132
   silver fulminate, 132
Fuming sulfuric acid. *See* Oleum
Fuses, 122
Fusible plug, 56

G

Gallium, 84
Gamma rays, 194-96
Gas cylinders
   fusible plug, 56
   safety-relief devices for, 55-56
      frangible disc, 56
      safety-relief valve, 56
Gas emergencies, 61
Gases
   air shipping of, 58
   cryogenic, 72-76
   compressed, 55-70
   highway shipping of, 58
   liquefied, 55
   medical, 58-59
   pressurized, 55
   rail shipping of, 56-57
   storage areas for, 58
Gas Law, 22-23
Gasoline, 97
Geiger counters, 198, 199
Gelatin dynamite, 130-31
Glottis, 169
Glycols, 101
Guide to Precautionary Labeling of
   Hazardous Chemicals, 206

H

Halocarbons, 51-52, 103-104
Halogen acids, 149-50
Halogens, 46, 51, 81, 90-91, 165
Handbook of Organic Industrial
   Solvents, 206
Hemoglobin molecule poisons, 167
Hexanitroethane, 126
Homologous series, 93

Hydrazine
   hazards of, 139-40
   use of, 139
Hydrides
   metal, 142
   nonmetal, 142-43
Hydrobromic acid, 149, 150
Hydrocarbon fires, 98-99
Hydrocarbons
   acetylene series, 95
   aromatic, 95-96
   defined, 93
   emergency action, factors
      determining, 97-98
   flammable liquids, 97
   petroleum products, 96
   saturated, 93-94
   unsaturated, 94-95. *See also*
      Substituted hydrocarbons
Hydrochloric acid, 149, 150
Hydrofluoric acid, 149
Hydrogen, 1
Hydrogen cyanide, 167
Hydroiodic acid, 149, 150
Hyperoxia, 164
Hypoxia, 164

I

Igniters, 122
Ignition temperature, 39
Indium, 84
Inorganic cyanides, 167
Ionization chamber, 199
Irritants, 159
Isobutyl alcohol, 101
Isocyantes, 116
Isopropyl alcohol, 101
Isotope, 2

J

Jaundice, 169

K

Ketones, 104-105
Kinetic-molecular theory
   chemical reactivity, 19
   heat and, 17-19
   liquids, evaporation and boiling
      of, 18-19
   physical states, 18
   pressure and, 17-18
   solids, melting of, 19

L

LD_{50}, 32
Liquefied gases, 55, 64-68
Liquid argon, 75
Liquid fluorine, 74
Liquid helium, 75
Liquid hydrogen, 75
Liquid krypton, 75
Liquid natural gas, 75
   leaking, 75-76
   storage facilities, 75
   transportation of, 76

Liquid neon, 75
Liquid nitrogen, 74-75
Liquid oxygen
    fighting fires involving, 73-74
    hazards of, 72-73
    precautions, 73
Liquid petroleum gases. *See* LP gas
Liquids, evaporation and boiling
        of, 18-19
Liquid xenon, 75
Liver poisons, 160, 176-81
    chemical compounds, 176
    heavy metals, 177-78
    organic chlorinated compounds,
        178-81
    toxicity limits, 181-82
LNG. *See* Liquid natural gas
Lower flammable limit, 35
Lower respiratory poisons
    acrolein, 167
    cyanides, 166-67
    sulfides, 167
LOX. *See* Liquid oxygen
LP fires, controlling, 67-68
LP gases, 65-68
    hazards of, 66-67
    leaking, handling, 67

M

Magnesium, 82
MAPP gas, 63-64
MCA Chem-Car Manual, 206
Medical gases, 58-59
Mercaptan, 60
Mercury cyanate. *See* Mercury
        fulminate
Metal hydrides, 142
Metallic elements
    fire fighting procedures, 84
    use, area of, 84
Metal peroxides, 135
Metals
    alkali metals, 80-81
    alkaline earth metals, 82-84
    as combustible materials, 79-85
    metallic elements, 84
    reactivity, factors affecting, 85
Methane, 3, 60-61
Methemoglobinaemia, 169
Mixtures, 4
MLD, 32
Molecular formulas, 8-9
Monomers, 109-10

N

Narcotics, 160
Natural gas, 60-61
Nerve impulse, 173
Nerve poisons, 160, 173
Neutron, 1
NFPA classification system, 39
Nitrating acid, 150
Nitric acid, 150, 185

Nitric ester explosives, 127-38
Nitrobenzene, 124
Nitrocellulose, 121
Nitro explosives, 124-26
Nitroglycerin, 121
Nonmetal hydrides, 142-43
Nonmetals
    carbon, 88
    halogens, 90-91
    phosphorus, 88-89
    sulfur, 90
Nucleus, 1
Nylon, 114

O

Official regulations for hazardous
        materials, 205-44
Oil of vitriol. *See* Sulfuric acid
Olefinic hydrocarbons, 94-95
Oleum, 149
Organic acids, 151
Organic bases, hazards of, 155-56
Organic peroxides, 136-37
Organic phosphorus compounds, 173
Oxygen, 42-43, 59-60

P

Perchloric acid, 150-51
Periodic law, the, 5-6
Peroxides, 112
    fire fighting procedures, 137
    hazardous properties, 136
    hydrogen peroxide, 135
    organic, 136-37
    reactions, 135-36
    storage, 136
    uses, 136
Phenol, 151
Phenolic, 114
Phosgene, 115, 166
Phosphine oxides. *See* Organic
        phosphorus compounds
Phosphoric acid, 185
Phosphorus, 88-89
    fire fighting procedure for, 89
    use, area of, 89
Physical changes, 4
Picric acid, 121
Plastics
    ABS, 113
    acrylics, 113
    cellulose, 113
    cellulose acetate, 113
    cellulose nitrate, 113
    epoxies, 113
    hazards common to, 116-17
    nylon, 114
    phenolic, 114
    polyamide-imides, 114-15
    polybutadiene, 115
    polycarbonates, 115
    polyesters, 114
    polyethylene, 115

Plastics (cont'd)
    polyimide, 115
    polymers, 112
    polypropylene, 115
    polystyrene, 115-16
    polyurethane, 116
    polyvinylchloride, 116
    silicones, 116
    thermoplastics, 109
    thermosetting plastics, 109
Platinum, 85
Poisons
    corrosive, 184-85
    liver, 176-81
    nerve, 173
    respiratory, 164-69
    systemic, 160
Polyamide-imides, 114-15
Polybutadiene, 115
Polycarbonates, 115
Polyesters, 114
    fluoroplastics, 114
Polyethylene, 115
Polyimide, 115
Polymer, 109, 112
Polypropylene, 115
Polystyrene, 115-16
Polyurethane, 116
Polyvinylchloride, 116
Portable tanks, 58
Positrons, 196-97
Pressure, 22-23
Pressurized gases, 55
Primacord, 132
Propane, 65-66
Propyl alcohol, 101
Pyrophoric, 42, 81
Pyrotechnics, 131-32
Pyroxylin plastic, 113

R

RAD. *See* Radiation absorbed dose
Radiation absorbed dose, 199
Radiation dose, 198
Radiation exposure, 199-201
Radiation flux, 198
Radiation injury, 201
Radiation intensity, 198-99
Radioactive material, transportation
        of, 201
Radioactivity
    alpha particles, 196
    beta rays, 196
    gamma rays and, 194-96
    hazards of, 194-201
    positrons, 196-97
    principles of, 189-92
Radioisotopes, 198-99
Radium, 82
Rale, 169
Rare gases, 75

Reactive materials
    carbides, 143-44
    hydrides, 142-43
    hydrazines, 139-40
    peroxides, 135-37
Red phosphorus, 89
REM. *See* Roentgen equivalent man
REP. *See* Roentgen equivalent physical
Resins. *See* Polymers
Resonance, 95
Respiratory poisons, 159, 164-69
    hemoglobin molecule poisons, 167
    inhalation problems, 168-69
    lower, 166-67
    symptoms, 169
    upper, 165
Roentgen, 198
Roentgen equivalent man, 199
Roentgen equivalent physical, 199

S

Safety-relief valve, 56
Scintillation counter, 198
Shock sensitivity, 122
Shock waves, 29-30
Silicones, 116
Simple asphyxiants, 159
Sodium chloride, 3, 50-51
Sodium cyanide, 167
Sodium hydroxide, 153-54
Solids, melting of, 19
Specific gravity, 40
Squibs, 122
Stoichiometric combustion
    conditions, 35

Straight dynamite, 130
Strontium, 82
Styrene, 116
Sublimation, 19
Substituted hydrocarbons
    acids, 105
    alcohols, 100-102
    aldehydes, 104-105
    amides, 105-106
    amines, 102-103
    esters, 105
    ethers, 106
    halocarbon, 103-104
    ketones, 104-105
Sucrose, 3-4
Suffocants, 159
Sulfides, the, 166
Sulfur, 90
Sulfuric acid, 149, 185
Sulfurous acid, 165
Survey meters, 198, 201
Systemic poisons, 160

T

Temperature, 22-23
Thallium, 84
Thermal sensitivity, 121
Thermoplastics, 36, 109
Thermosetting plastics, 109
Threshold limit, 32
TL. *See* Threshold limit
TNT, 124
Toxicity, 31-33
Toxicity limits, 161
    liver poisons, 181-82

Toxic materials
    absorption of, 160-61
    anesthetics, 160
    asphyxiants, 159
    irritants, 159
    narcotics, 160
    systemic poison, 160
Toxicology, principles of, 159-61
Trinitrochlorobenzene, 125
Trinitrotoluene, 121
Tritium, 2
Tube trailers, 58

U

United States Department of Trans-
    portation, 181
Unsaturated polyesters, 114
Upper flammable limit, 35
Upper respiratory poisons
    dioxides, the, 165
    halogens, 165
Urethanes, 116

V

Valence, 8
Vapor density, 40
Vapor pressure, 18
Violet phosphorus, 89
Volume, gas laws governing, 22-23

W

Water, reactivity with, 45-47
Water solubility, 39-40
White phosphorus, 88-89

X

X-rays. *See* Gamma rays